Reliability-Centered
Maintenance

Other McGraw-Hill Books of Interest

Reliability-Centered Maintenance

Anthony M. Smith, P.E.

McGraw-Hill, Inc.

New York   San Francisco   Washington, D.C.   Auckland   Bogotá
Caracas   Lisbon   London   Madrid   Mexico City   Milan
Montreal   New Delhi   San Juan   Singapore
Sydney   Tokyo   Toronto

Library of Congress Cataloging-in-Publication Data

Smith, Anthony M.
  Reliability-centered maintenance / Anthony M. Smith.
    p. cm.
  Includes index.
  ISBN 0-07-059046-X
  1. Plant maintenance. 2. Reliability (Engineering).
3. Maintainability (Engineering). I. Title.
TS192.566 1993   92-47464
658.2'02—dc20   CIP

2 3 4 5 6 7 8 9 0 DOC/DOC 9 8 7 6 5 4

ISBN 0-07-059046-X

The sponsoring editor for this book was Gail F. Nalven, the editing supervisor was Christine Furry, and the production supervisor was Suzanne W. Babeuf. It was set in Century Schoolbook by North Market Street Graphics.

*To Mary Lou and the crew:
you are the reason for all of this.*

Contents

Foreword

There are four Ts that are absolutely essential for your employees to provide error-free work. They are:

1. Training
2. Tools
3. Time
4. Teams

Training is essential to prepare the employees to do the job right every time. Well-maintained tools provide the means for the employee to excel. Time is necessary to ensure that the job is done right every time. A support team that works like a well-oiled clock is needed to make everything come together at the right instant and place. All four of these Ts are required to reach the level of excellence needed to compete in today's challenging worldwide marketplace. Management's job is to provide these four Ts and, when they do, they are amazed at what the employees can accomplish. Meshing these four Ts provides the organization with a jump start in becoming a world-class operation. Take away any one of them, and you immediately place your organization in a noncompetitive position.

In the 1980s, American business leaders began to get serious about improving their competitive position. Their first focus was on training—training to get people to work together to solve problems and to perform their present jobs even better. Management focused their attention on reducing the scrap and rework cost. Many organizations eliminated production standards to encourage the employee to take the time to do it right every time. The result was a reduction in product cost, since scrap and rework cost fell much more than the first-time process cost increased. Teams became fashionable. Many organizations measured the success of their quality effort by the number of teams they formed or the percentage of employees trained in team methods.

The result was a major improvement in quality, but it did not stop the slow and constant loss of our market in key manufacturing products like thin film devices, automobiles, televisions, computers, tires, etc. Low-tech products have moved out because of manufacturing costs. High-tech companies are losing market share to overseas competitors because management has not made good use of the limited resources dedicated to:

1. Acquiring and maintaining state-of-the-art equipment
2. Developing advanced techniques
3. Developing new products

Today's competitive battle is not being fought on the factory floor—we are losing the competitive battle in the support areas. This is the reason that this book, *Reliability-Centered Maintenance,* is so important to both high-tech and low-tech companies.

To succeed in manufacturing today, you must have the right equipment, and it must be superbly maintained. The tendency in the United States is to cut direct labor forces and reduce variability by automating and mechanizing as many operations as possible. This trend has turned even the low-end product manufacturing facility into a high-tech production environment. As computerized, complex equipment replaces the human, a major burden has been placed upon the organization's maintenance personnel. To keep a trained work force of maintenance personnel who are familiar with the complex, electromechanical equipment and the software required to support it is one of the biggest challenges many companies face today. Add to that the cost of a warehouse full of expensive spare parts and the problem becomes overwhelming. The result is that the maintenance budget is becoming a bigger and bigger part of the total operations budget. In some high-tech companies, maintenance costs are exceeding direct labor costs.

As a result, management has aggressively put the brakes on what they view as a runaway freight train. This has forced the maintenance organizations to place their priority on reactive measures (aimed at restoring failed equipment to service) at the sacrifice of proactive measures (aimed at preventing, retarding, or mitigating failures in the first place). One machine operator confided to me that "The maintenance people around here are so busy fixing failures that they never have time to work on anything that isn't broken down. So, when my machine starts to act up, I break it so that it is unusable. That gets the priority I need to get it fixed. If I turn in a trouble call, I would be told to keep on running my machine and measure all the parts to be sure that no bad ones get out. You know what that does to my performance record, don't you?"

There are many problems with maintenance programs today. Some equipment gets too little, while some gets too much, preventive maintenance. Operators are not trained to understand how the equipment works. They are trained to know how to run it, but that is not enough. Maintenance people do not take time enough to truly understand why things fail. A motor fails. It is replaced. It fails again 30 days later, and it is replaced again. Maintenance does not have or take the time to analyze the situation to understand that the bad bearing in the pulley shaft is causing excess torque when the bearing gets hot, causing the motor to burn out. Maintenance reporting systems as a whole are poor and do not provide the data required to diagnose complex problems. The maintenance job has become very complex, compared to 20 years ago. The complex interface existing in much of the equipment requires a knowledge of the mechanics, programming, and electronics that would challenge any three engineers. With such a complex problem facing management, one proven approach to solving these seemingly insurmountable problems rests with a relatively new approach to maintenance called *reliability-centered maintenance*. After years of wrestling with the problem first-hand, Mac Smith has taken the reliability-centered maintenance concepts so successfully developed and employed by the commercial aviation industry, and has shown how to apply them with equal success in industrial plants and facilities.

I believe that reading this book and implementing the concepts presented will provide you with that extra competitive edge that can set you apart from the rest. I hope you implement it before your competition does.

Dr. H. James Harrington
The International Quality Advisor
Ernst & Young

Preface

There are occasions in a professional career when a series of seemingly unrelated events can suddenly align to produce an unbelievably creative situation. In my experience, such occasions are few and far between, and for some they might never occur. The seeds which led to this book came from one of those rare alignments that happened to me.

The first segment of this alignment I owe to a career of 24 years with General Electric, and an association over those years with a multitude of peers and bosses who helped me to develop an appreciation of the product design and development process and, most importantly, the necessity to always be sensitive to the fact that eventually someone would have to build that product, and then use (operate) it. During my years at GE, I was frequently given the opportunity to personally experience the user environment and the impact of product design on the user. That impact was not always totally positive—you remember these experiences and try to learn from them. My exposure was mainly in the area of aerospace systems, jet engines, and power generation plants. While much of my early engineering career was directly in the area of product design and development, a later and extensive association with the reliability and maintainability disciplines was the principal path that led to my frequent dealings with the manufacturing and operating phases of our products and systems. To say the least, I became acutely aware of the fact that the world of product design and development is vastly different from the world in which those products are operated. In a certain sense, the product design/development world, while challenging and exciting, is a rather safe and somewhat sterile environment. It is generally conducted in a clean, temperature controlled, and friendly environment. The resources necessary to do the job are at hand or readily available. If you don't get the job done today, chances are that tomorrow is acceptable. Not so in the operational world! The physical environment where the work is done is often hot, humid, dusty, and sometimes downright dangerous. You may or may not always have exactly the right resources to keep the system operating—yet you may not have the luxury of waiting until tomorrow to restore a failed system to operation. Unless safety of the customer and/or operator is at risk, you often find that a great deal of pressure builds to get the system back on-line. And your ability to do just that is often greatly influenced by how well the product designers accounted for the environment in which you operate. As a simple example, did the designer leave you enough room to remove and replace a failed black box, or must you do near impossible handstands with special tools to get it done? Over the years, I was fortunate to have the exposures and experiences from which a sensitivity to the needs and conditions of the operational world could be developed.

The second segment of this alignment came after I had gone into the consulting business, where I continued to work with the operating environment. This segment developed as the direct result of a very fortuitous professional friendship that evolved over several years with a remarkable engineer—by name, Thomas D. Matteson. Tom, who retired as Vice President of Maintenance Planning for United Airlines, was essentially the major innovator and force behind the detailed development of the methodology used by a Maintenance Steering Group (MSG-1) to produce the preventive maintenance program for the 747 aircraft. This methodology was so successful that, subsequently, all modern jet aircraft have employed the MSG-1 approach to define their preventive maintenance programs for the Federal Aviation Administration (FAA) Type Certification process. MSG was later anointed with the title Reliability-Centered Maintenance, or RCM, when United Airlines was contracted by the Department of Defense to write the first (and until now, only) textbook on RCM in 1978 (see Reference 6). Due to our mutual interest in the regulatory process (Tom's with the Federal Aviation Administration and mine with the Nuclear Regulatory Commission), we would frequently discuss various reliability, safety, operations, and maintenance problems that had a surprising degree of procedural similarity. Also, by 1980, U.S. commercial aircraft had accumulated about an order of magnitude more operating hours than U.S. commercial nuclear reactors and were thus a considerably more mature product. It became apparent in our discussions that nuclear plants were experiencing many of the maturing pains that commercial aircraft had passed through years before. Thus, in 1981, Tom and I teamed to take our mutual interests in reliability, safety, operations, and maintenance to the Electric Power Research Institute (EPRI) with the suggestion that nuclear utilities could potentially benefit from several of the proven practices and procedures that had evolved over the years in the commercial aviation industry. The ensuing EPRI-sponsored study (see Reference 21) was a very successful endeavor in that it led directly to a utility interest in RCM, and later the conduct of the first two RCM pilot projects at the Turkey Point and McGuire Nuclear Power Plants of Florida Power and Light and Duke Power, respectively (see References 13 and 14). The story from there is one in which RCM has become a preventive maintenance (PM) methodology of interest across a broad spectrum of the utility industry (both nuclear and fossil) and, most recently, has seen a great deal of interest developing in other industries as diverse as oil refineries and railroads.

Thus, the fortuitous alignment of my sensitivity to the user environment with the almost accidental chance to work with the "father of RCM" created for me the opportunity to work with and introduce RCM to other segments of industry. I learned RCM from Tom Matteson. He was a great teacher. I am indebted to him for his patience and tenacity in helping to make me an RCM expert.

So why did I decide to write this book? There are two reasons:

First, after Tom and I completed the first two RCM pilot projects at the Turkey Point and McGuire Nuclear Power Plants, there was a rush of several consulting firms into the RCM picture because of the understandable business opportunity that it presented. From my point of view, several of these firms tried to make the RCM process "better" by introducing some fundamental changes to the process, and by adding some new bells and whistles, all of which only increased cost without any significant added value.

Second, there is no generally available textbook on the market (to my knowledge) which presents a detailed discussion of the classical RCM process as it applies to non-aircraft systems.

It is my hope that this book will help to solve both of these situations, and permit me to leave a legacy to the operations and maintenance (O&M) world that will foster a continued accrual of long-term benefits that can be realized from a proper RCM application.

Anthony M. Smith
Saratoga, California
July 1992

Acknowledgments

No man is an island. I am both humbled and awed by the support, assistance, and information that I have received in putting this book together for you. The reasons for this book, and the origins of my entry into the world of RCM, are the subject of the Preface which I encourage you to read, if you have not already done so.

My experiences with a broad cross section of industry are the real reasons why I have been able to put much of this material together. Thus I would first like to acknowledge the companies which form that background of experience:

Duke Power—McGuire Nuclear Plant

Electric Power Research Institute—Palo Alto, California

Ernst & Young—San Jose, California

Florida Power & Light—Turkey Point and St. Lucie Nuclear Plants
                    —Port Everglades and Martin Fossil Plants

GPU Nuclear—Three Mile Island-1 and Oyster Creek Nuclear Plants

Illinois Power—Clinton Nuclear Plant

Labatts Brewery—Edmonton, Canada

Niagara Mohawk Power—Nine Mile Point Nuclear Plants

Mississippi Power—Watson Fossil Plant

Sandia National Laboratory—Solar One Plant

Shell Oil—Houston, Texas

Southern Pacific Transportation—San Francisco, California

Sverdrup Technology—USAF Engine Test Facility

Washington Public Power Supply System—WNP-2 Nuclear Plant

Westinghouse Electric—Bellingham and Sayreville Combined Cycle Plants

Beyond that, I have received invaluable assistance from certain organizations and, more particularly, specific individuals therein, in terms of permission to use their RCM program material as well as a review and critique of the book for its accuracy and clarity, plus suggestions for meaningful additions. I am especially indebted to these people, whom I sincerely thank:

Ramesh Gulati—Sverdrup Technology/U.S. Air Force

H. James Harrington—Ernst & Young

Glenn R. Hinchcliffe—Florida Power & Light

M. G. (Pete) Snyder—GPU Nuclear/TMI-1

David H. Worledge—Electric Power Research Institute

Ron Worthy—Westinghouse Electric

It takes a world of skill and patience to do all of the word processing of the draft material, and my thanks go to Lorabeth Brink for that consistent support. Finally, I am indebted to the staff at McGraw-Hill, especially Gail Nalven, for their efforts in accepting this manuscript and producing it in such a professional manner.

Reliability-Centered
Maintenance

The Preventive Maintenance (PM) Process—Opportunity and Challenge

Until recently, product development and manufacturing engineering have been the dominant technical disciplines in the U.S. industrial community, with operations and maintenance (O&M) occupying a back seat in the priority of corporate success strategies. The past decade has seen this picture shift rather dramatically to where O&M is now a peer with the development and manufacturing disciplines. There are compelling reasons for this, not the least of which is the decisive role that O&M now plays in issues ranging from safety, liability, and environmental factors to bottom line profitability. With O&M now center stage, Preventive Maintenance (PM) optimization is providing never-before-seen opportunities and challenges to the O&M specialists.

Some of these challenges come in the form of various maintenance problems that currently have a great deal of commonality across our industrial system. Ten such problems are briefly discussed to indicate some specific dimensions on the challenge before us.

Many share the view that the Reliability-Centered Maintenance methodology offers the best available strategy for PM optimization. The author strongly shares this view. And that is what this book is all about.

1.1 The Current Maintenance Picture

The title of this introductory chapter contains two key words that set the stage for the continuing theme throughout this book—challenge and opportunity. Let's step back for a moment from the everyday pressures and excitement that are typically associated with a plant or facility operation, and look at what these words might mean to us.

Since the end of World War II, the growth of the U.S. industrial complex has been dominated by two factors: (1) technical innovations which have led to a plethora of products that were mere dreams in the pre-World War II period and (2) volume production which has enabled us to reach millions of customers at prices well within the reach of virtually every consumer in the United States and its international peers. From a motivational point of view, product development and design, as well as the manufacturing engineering that provides for mass production capabilities, have been the "darlings" of the engineering world for the past 30–35 years. And, as a result, the bad news is that operations and maintenance (O&M) have often been relegated to a "necessary evil" role with all of the attendant problems of second-class citizen status when it comes to research and development (R&D) projects, budget requests, manpower allocations, and management awareness. Somehow, the reasoning goes, those good people in the trenches seem to keep things running—so they get the token pat-on-the-back and another year's supply of bailing wire and chewing gum to keep it all together. A bit of an exaggeration? Maybe, but some of you are probably relating to this scenario in some fashion.

Well, here is the good news. Things have been rapidly changing over the past decade, as you have most likely noticed. The reasons for this are legend—environmental concerns, safety issues, warranty and liability factors, regulatory matters, and the like. But most of all, as plant and equipment ages, and global competition becomes a way of life, management has realized that O&M costs are (or could be) "eating their bottom line lunch." There has been a dramatic positive shift in management concern and awareness about both the cost and technical innovation of O&M policies, practices, and procedures. In fact, we might even go so far as to say that O&M now occupies a peer position with product development and manufacturing engineering in many product areas.

And that is where the challenge and opportunity comes into this picture. Make no mistake, O&M is now in the center-stage spotlight. What are we going to do about it?

It is hoped that this book will help you to answer that question with regard to one major aspect of the O&M challenge—how to get the most from the resources committed to the plant or facility preventive maintenance program. It is suggested that a viable strategy can be PM optimization via the use of the Reliability-Centered Maintenance methodology. And that is what this book is all about.

1.2 Some Common Maintenance Problems*

With the O&M spotlight coming center stage, it is instructive to look at some industrywide maintenance history of the past two decades, especially with respect to some of the more classic maintenance problems that we need to address. It is recognized that the list of problems discussed here is not all-inclusive, nor are these problems necessarily common to everyone. But they are thought to represent a mainstream of experiences that have been observed with sufficient frequency to warrant their attention here.

1. *Insufficient proactive maintenance.* This problem clearly heads the list simply because the largest expenditure of maintenance resources in plants typically occurs in the area of *corrective* maintenance. Stated differently, the vast majority of plant maintenance staffs operate in a reactive mode, and in some instances plant management actually has a deliberate philosophy to operate in such a fashion. It is interesting to note that, in the latter case, the end product from such a plant usually has the highest unit cost within the group of peers producing the same product. A major contributor to the unit cost thus accrues from a combination of the high cost to restore plant equipment to an operable condition coupled with the penalty associated with lost production. Since simple arithmetic readily demonstrates this fact, it is really quite surprising that this reactive environment continues to occur. Yet, to one degree or another, it seems to be a commonplace situation.

2. *Frequent problem repetition.* This problem, of course, ties in rather directly with the preceding. When the plant modus operandi is reactive, there is only time to restore operability. But there is never enough time to know why the equipment failed, let alone enough time or information to know how to correct the deficiency permanently. The result is that the same problem keeps coming up—over and over. This repetitive failure problem is often discussed in terms of root cause analysis, or more appropriately the lack thereof. Unless we understand why the equipment failed and act to remove the root cause, restoration may be a temporary measure at best.

3. *Erroneous maintenance work.* Humans make mistakes, and errors will occur in maintenance activities (both preventive and corrective). But what is a tolerable level of human error in a maintenance program? Is it 1 error in 100 tasks, 1 in 1000, 1 in 10? The answer could depend on the consequence realized from the error. If you are a frequent flyer, you would like to believe that both maintenance errors and

---

* Inputs provided by the Electric Power Research Institute and the EPRI RCM Users Group are gratefully acknowledged as source material for this discussion.

pilot errors are less than 1 in 1 million! (In terms of catastrophic errors, they are in this range.) But let's think about this in economic terms—the cost of another corrective action and attendant loss of plant production. Most plant managers wish to have that 1 in 1 million statistic, but seem to believe that reality is more like 1 in 100. There is strong evidence, however, that human error is the cause of more than 50 percent of plant forced outages, and that some form of human error might be occurring in some locations in one of every two maintenance tasks that are performed. Surprised? Check your own records. You may be in for a shock.

4. *Sound maintenance practices not institutionalized.* One way to solve the human error problem is, for starters, to know the practices and procedures which can assure that mistakes are not made—and then to institutionalize them in the everyday work habits at the plant. Collectively, industry has a great deal of knowledge and experience on how equipment should be handled (e.g., how it should be removed from the plant, torn down, overhauled, reassembled, and reinstalled). Individual plants are usually informed on only a small percentage of this collective picture. Worse yet, what is known is all too infrequently committed to a formalized process (procedures, training, etc.).

5. *Unnecessary and conservative PM.* At first glance, one might feel that this problem is in conflict with item 1. While the need for more PM coverage is clearly an appropriate issue, there is a parallel need to look at the PM which is currently performed in terms of "Is it right?" Historical evidence strongly suggests that some of our current PM activities are, in fact, not right. In some cases, PM tasks are totally unnecessary because they have little, if any, relationship to keeping the plant operative. (In later chapters, we will see that this problem relates to the lack of what we call task "applicability".) It is not uncommon to examine a plant PM program, and find that 5 to 10 percent of the existing tasks could be discarded and the plant would never know the difference. Trouble is that most plants never revisit PM tasks with this question in mind. A second form of this problem is one wherein the PM task is right, but too conservative. This problem is usually associated with task frequency (i.e., the frequency requires the PM action too often). This seems to be especially true of major overhaul tasks where there is some substantial evidence to suggest that 50 percent or more of the PM overhaul actions are performed prematurely.

6. *Sketchy rationale for PM actions.* Did you ever ask the maintenance manager why some PM task is being performed? Did you receive a credible response? Could the response be supported with any documentation that could reasonably reconstruct the origins of the task (other than "the vendor told us to do it"—see item 8 following)? Unfortunately, the absence of information on PM task origin or any docu-

mentation to clearly trace the basis for plant PM tasks is the rule, and not the exception. Perhaps one might suggest that this is not all that unacceptable. If maintenance costs (PM + CM) were low and still decreasing, and if plant forced outages were virtually nonexistent, one might allow that this could be the case. But neither of these factors are so outstanding that we can continue to ignore the issue of why we do a PM task, nor to forego the ability to record the basis for such actions in appropriate documentation. For example, the Federal Aviation Administration requires an approved and documented basis for PM as a requirement for aircraft Type Certification (i.e., approval to build and sell the airplanes), a requirement that has been in place for several decades. More recently, nuclear power plants have also been evaluating the necessity of establishing a more formalized documentation process as one element of their response to the new Maintenance Rule that was issued by the Nuclear Regulatory Commission.

7. *Maintenance program lacks traceability/visibility.* If the plant does not perform root cause analysis on equipment failures, and is remiss at recording the basis of PM actions, then at least two significant areas have been defined where visibility and traceability of decisions/actions are missing. But the problem goes beyond this in many situations, and we refer here to the lack of any definitive Maintenance Management Information System—the MMIS. Frequently, there is no traceable record of PM actions and costs to be found anywhere except in the heads or desk drawers of the plant staff. If they leave, the plant memory walks out the door with them. In today's world of modern and inexpensive computer systems, complete with necessary software, there seems to be little excuse not to have good plant records on what was done (or is scheduled to be done), and why management decided on certain definitive strategic and tactical courses of action.

8. *Blind acceptance of OEM inputs.* The Original Equipment Manufacturer (OEM) almost always provides some form of operations and maintenance manual with the delivered equipment. From a PM point of view, two problems develop with this input. First, the OEM has not necessarily thought through the question of PM for the equipment in a comprehensive and cost-effective fashion. Often, the OEM PM recommendations are last-minute thoughts that tend to be aimed at protecting the manufacturer in the area of equipment warranty (this is the origin of many conservative PM tasks). Second, the OEM sells equipment to several customers, and these customers operate that equipment in a variety of different applications—for example, cyclic rather than steady state, very humid rather than dry ambient conditions, etc. The OEM usually designs the equipment with some operational variability in mind, but rarely does the vendor ever specifically tailor the equipment to your special needs. Yet the basis for many PM programs is the

blind acceptance of OEM PM recommendations as the best course of action to pursue—even though the OEM recommendations are conservative and not necessarily applicable to the plant's operating profile.

9. *PM variability between like/similar units.* Within a given company, it is likely that multiple plant locations are involved in production and, in some instances, each plant may even have multiple units at each location. The utility industry typifies the latter situation where two or more power generation units are frequently located at each plant site. These multiple plant or unit situations are likewise composed of production facilities that are often virtually identical from site to site or, at the very least, contain a wide spectrum of equipment that is identical or highly similar. Under these circumstances, it would seem reasonable to assume that their PM programs share this commonalty in order to standardize procedures, training, spare inventories, etc. to capitalize on the obvious cost savings that can be achieved. Unfortunately, this is not a good assumption; more often than not we find that each plant location tends to be its own separate entity with many of its O&M characteristics different from its sister plants within the company. It is not clear why corporate management allows this to occur, but at the plant level of organization there appears to be a strong feeling of "I'm not like them" and "We know what's best for us" attitudes driving this lack of commonality. Across a given industry, of course, this lack of commonality becomes even more pronounced (and perhaps a bit more understandable).

10. *Paucity of predictive maintenance applications.* There is an entirely new area of maintenance technology that has been developing for several years which is usually described under the name of *predictive maintenance.* It is also described with titles such as *condition monitoring, monitoring and diagnostics,* and *performance monitoring.* All of these names are intended to describe a process whereby some parameter is measured in a nonintrusive manner and trended over time, said parameter being one with a direct relationship to equipment health, or at least some specific aspect of equipment health. Clearly, this process has the potential for significant payoff when it can be used to tell us when it is necessary to perform some maintenance task, thus precluding both unnecessary as well as premature preventive maintenance actions that otherwise would occur. Some of this technology is fairly sophisticated (e.g., vibration analysis on rotating machinery) and some of it is fairly simple (e.g., pressure drop across a filter). But, to a large extent, much of this technology has not been introduced into our plants and facilities. And, where a plant has a predictive maintenance program, more often than not, its focus is on the deployment of the sophisticated, not the simple, technology.

1.3 Where to from Here?

All of these 10 problem areas that were selected for inclusion here have solutions that can be applied in your plant or facility. It is not the intention to discuss specific solutions at this point, but it is important to understand that none of these 10 problems (plus others that you could likely add to the list) represents insurmountable difficulties. The solutions that can be employed will, however, require a combination of available engineering and technical skills, as well as management awareness and motivation. Let me say that I have searched for several years now to find an appropriate methodology that could be employed to develop strategies and programmatic approaches to handle such problems as those listed previously. And let me further say that this search has convinced me that the Reliability-Centered Maintenance methodology offers the most systematic and efficient process that one can find to address an overall programmatic approach to the optimization of plant and equipment maintenance. Hopefully, this book will tell you why and how to use RCM to achieve such results.

The book has been structured such that Chaps. 1–3 deal with a general discussion of issues and problems related to current maintenance practices and procedures, as well as a discussion of reliability principles as they relate to maintenance. Chapter 4 then introduces the RCM methodology. In Chaps. 5–8, we deal with the specific details of the RCM process, a variety of RCM examples, and the business of carrying an RCM program to on-the-floor implementation. Let me suggest that you may find it worthwhile to *initially scan Chap. 8 before proceeding through the book* in order to capture the flavor of what it takes to successfully implement an RCM program. A quick background from the Chap. 8 material will help you to understand more clearly why certain points are emphasized throughout Chaps. 2–7.

Finally, a large segment of corporate America is intensely pursuing a variety of programs that are aimed at accomplishing quantum jumps in the productivity and quality of their goods and services. The competition that we have experienced from the Japanese is foremost in motivating an awareness of the fierce global competition that we face. The programs most often employed to achieve these quantum improvements are forms of the Total Quality Management (TQM) process that is used by the Japanese as a way of life in their industrial environment. For those readers interested in the TQM process, App. A is included to discuss both the TQM process and the role that preventive maintenance (and thus RCM) can play in the TQM strategy.

Preventive Maintenance Definition and Structure

In Chap. 2, we develop several basic elements of a preventive maintenance program. Initially, we define preventive maintenance (as distinct from corrective maintenance which is often a source of confusion), and delineate why preventive maintenance is performed (which many people tend to view too narrowly). This leads then to a discussion of four major PM task categories that can be employed. A logical process for formulating a PM program is suggested followed by the author's views and experiences on the current practices and myths that are employed to specify equipment PM tasks. Finally, we examine some of the key disciplines, both management and technical in nature, that can and should be used in supporting PM programs.

2.1 What Is Preventive Maintenance?

At first glance, to ask "What is preventive maintenance?" seems to be a rather mundane, if not totally unnecessary, question to pose. However, experience has clearly shown that some confusion does exist over just what people mean when they use the term *preventive maintenance*. There are a variety of possible reasons for this confusion. One significant factor stems from the evidence that a vast majority of our industrial plants and facilities have been operating for extended periods, years in many cases, in a *reactive* maintenance mode. That is to say that their maintenance resources have been almost totally committed to responding to unexpected equipment failures and very little is done in the preventive arena. Corrective, not preventive, maintenance is frequently the operational mode of the day, and this tends to blur how many people view what is preventive and what is corrective.

In an extreme case, a plant can develop an entire culture that fosters a feeling of pride in their ability to fix things rapidly and under pressure when a forced outage occurs—and rewards are consistently given for such performance. The operating philosophy under these conditions is almost totally reactive and corrective in nature, but the plant personnel tend to view their actions as preventive in the sense that they were able to "prevent" a long outage because of their highly efficient and effective reactive and corrective actions. Throughout this book, we shall use the following definition of preventive maintenance (PM):

> *Preventive maintenance* is the performance of inspection and/or servicing tasks that have been *preplanned* (i.e., scheduled) for accomplishment at specific points in time to retain the functional capabilities of operating equipment or systems.

The word *preplanned* is the most important word in the definition; it is the key element in developing a *proactive* maintenance mode and culture. In fact, this now provides us with a very clear and concise way to define corrective maintenance (CM):

> *Corrective maintenance* is the performance of *unplanned* (i.e., unexpected) maintenance tasks to restore the functional capabilities of failed or malfunctioning equipment or systems.

People do play games with the terminology and its interpretation. These games can be driven by such diverse nontechnical factors as accounting practices or political (regulatory) pressures. For example, some plants, in addition to planned outages for major preventive maintenance tasks and forced outages for unexpected failures with resultant shutdown or cutback in operations, have a third category known as a *maintenance outage* (MO). The MO occurs as a result of an *unexpected* equipment problem which hasn't quite yet reached the full failure state—but will do so very soon (e.g., within hours or a few days). So the plant management will delay the shutdown or cutback until some off-peak period when the plant outage is more tolerable, and hope that the equipment will hold out until then. Now from an operational point of view, this is a very smart thing to do—but as a rule, MOs are *not* counted when it comes to reporting the plant forced outage rate. Somehow they seem to wind up in the preplanned category ("after all, we planned to fix it next Saturday!"). Make no mistake about it, an MO is a forced outage and should be labeled as such when measurements are made. You are only kidding yourself to do otherwise.

As a general rule, corrective maintenance is more costly than preventive maintenance. As the man in the Pennzoil ad says, "Pay me now or pay me (more) later." And of course we all know the old saying, "An ounce of prevention is worth a pound of cure." These catchy phrases are not just idle conversation pieces. They come from the experience of hard knocks. If anyone should doubt this, then just compare two similar plants or systems where one has a proactive maintenance program and the other a reactive maintenance program. Which one do you think has the lower overall maintenance cost and higher availability?

In later chapters, we will find that the use of corrective maintenance in a proactive maintenance program is occasionally a very viable option *if* it is done under very carefully defined and controlled circumstances, which we will spell out in detail in Chap. 5. But the general rule that corrective maintenance should be avoided in favor of preventive maintenance is still the proper way to think.

2.2   Why Do Preventive Maintenance?

This question, too, appears on the surface to be mundane—perhaps even unnecessary. But here is why we consider it important to raise this question early in the book. We find that the popular belief about why PM should be done is rather narrowly defined and, as such, leads to the exclusion of a number of golden opportunities for PM enhancement.

So why do *you* do preventive maintenance? The overwhelming majority of maintenance and plant engineering personnel will respond, "To prevent equipment failures." Would that have been your response? If so, you are correct—but not complete in your viewpoint. Unfortunately, we are not yet smart enough to prevent all equipment failures. But that does not mean that our ability to perform meaningful preventive maintenance tasks must end there. In fact, there are two additional and important options to consider. First, while we may not know how to prevent a failure, frequently we do know how to detect the onset of failure. And our knowledge of how to do this is increasing every day, and is creating a whole new discipline called predictive maintenance. Second, even though we may not be able to prevent or detect the onset of failure, we often can check to see if a failure has occurred before an equipment is called into service. Various standby and special purpose equipments (whose operational state is often hidden from view until it is too late) are candidates for this area. Thus, discovery of hidden failures is the third PM option available to us.

To summarize, there are *three* reasons for doing preventive maintenance:

1. prevent failure

2. detect onset of failure

3. discover a hidden failure

2.3 Preventive Maintenance
Task Categories

By identifying the three reasons for doing preventive maintenance, we have also set the stage for defining three of the four task categories from which a PM action may be specified. These task categories, by one name or another, are universally employed in constructing a PM program, irrespective of the methodology that is used to decide what PM should be done in the program. The four task categories are as follows:

1. *Time-directed (TD)*   aimed directly at failure prevention or retardation.
2. *Condition-directed (CD)*   aimed at detecting the onset of a failure or failure symptom.
3. *Failure-finding (FF)*   aimed at discovering a hidden failure before an operational demand.
4. *Run-to-failure (RTF)*   a deliberate decision to run to failure because the others are not possible or the economics are less favorable.

Each of these will be discussed in more detail to clarify what they cover and how they might be used.

Time-directed (TD).   In the not too distant past, virtually all preventive maintenance was premised on the basis that equipment could be periodically restored to like-new condition several times before it was necessary to discard it for a new (or improved) item. This premise thus dictated that equipment overhauls were about the only way to do preventive maintenance. Thus at specified "hard time" intervals, overhauls were done regardless of any other consideration. Today, we are slowly realizing that this is not always the correct path to pursue. However, in many valid situations we still specify PM tasks at predetermined ("hard time") intervals with the objective of directly preventing or retarding a failure. Conditions under which this approach is valid are discussed in Chap. 5. When such is done, we call it a *time-directed* or TD task. A TD task is still basically an overhaul action—sometimes very complete, extensive, and expensive (like rebuilding an electric motor), and sometimes very simple and cheap (like alignments and oil/filter replacements). As a rule of thumb, whenever we have a planned intrusion into the equipment (even just to inspect it), we have in essence an overhaul-type action and TD task.

The keys to categorizing a task as time-directed are: (1) the task action and its periodicity are preset and will occur without any further input when the preset time occurs, (2) the action is known to directly provide failure prevention or retardation benefits, and (3) the task action requires some form of intrusion into the equipment.

Condition-directed (CD). When we do not know how to directly prevent or retard equipment failure—or it is impossible to do so—the next best thing that we can hope to do is to detect its onset and predict the point in time where failure is likely to occur in the future. We do this by measuring some parameter over time where it has been established that the parameter correlates with incipient failure conditions. When such is done, we call it a *condition-directed* or CD task. Thus, a CD task would prewarn us to take action to avoid the full failure event. If the warning comes soon enough, our action can most likely be taken at some favorable timing of our choice. Note that the CD task is dramatically different than the MO situation in that our knowledge of failure onset is a deliberate and preplanned input, as is the action to be taken when the failure onset is detected; the MO is the result of a totally unplanned occurrence. The CD task, like the TD task, has a periodicity for the measurements, but actual preventive actions are not taken until the incipient failure signal is given. The CD task takes two forms: (1) we can measure a performance parameter directly (e.g., temperature, thickness) and correlate its change over time with failure onset or (2) we can use external or ancillary means to measure equipment status for the same purpose (e.g., oil analysis or vibration monitoring). With the CD task, all such measurements are nonintrusive. The keys to classifying a task as CD are: (1) we can identify a measurable parameter that correlates with failure onset, (2) we can also specify a value of that parameter when action may be taken before full failure occurs, and (3) the task action is nonintrusive with respect to the equipment. Note that if the parameter behaves in a stepwise fashion, as is often the case with digital electronics, it is probably of no use for a CD task.

Let's take a couple of examples to illustrate the CD task. First, we'll look at a rather simple situation that we all encounter at one time or another—the automobile tire. This is an especially interesting example to review because it illustrates several important points. The CD task that is employed here (as a rule, somewhat informally) is a performance monitoring of tire tread thickness. We periodically inspect the tires, or the dealer service department does so at predetermined PM shop visits, and when the tread thickness reaches $\frac{1}{32}$ in, we get new tires. Notice that tire manufacturers do *not* recommend an automatic replacement at, say, 25,000 miles (a time-directed task) because they cannot accurately predict the proper mileage for such action due to the many variables influencing wear. Notice that they do, however, recommend the performance of TD tasks such as wheel alignment and balance at prescribed fixed intervals in order to help us to get the maximum possible life from the tire. But no amount of PM will prevent eventual tire failure (if left unnoticed) during the useful life of the automobile. With a little added sophistication in the CD task, we can record tread thickness as a function of miles traveled for our specific usage habits, and actually predict when replacement is likely to be needed. If we need to plan for such an investment, as is frequently the case with large truck fleets, this prediction information can prove to be invaluable.

A second, and more technically complex, example might be the use of oil analysis on jet engines, where we measure for chemical and solid contaminants as indicators of wearout and/or incipient failure conditions in hidden parts within the engine; or the use of vibration monitoring sensors on rotating shafts where known limits on shaft movement will be breached when bearing failure onset is developing.

Notice that in the tire example, our knowledge may never be sufficient to specify a TD task for tire replacement (too many variables with too many people involved). But in the case of the oil analysis and vibration monitoring, it is very likely that our knowledge of the failure mechanisms and causes involved will someday be well understood—and these CD tasks will be replaced with TD tasks. As a general rule, it can be said that our current knowledge of failure mechanisms and causes is rather sparse (but we do a lot of guessing anyway); thus, the potential for CD application is large. As our knowledge base increases, we should see the gradual shift from CD to TD tasks. This shift will be a long-term evolutionary process.

Failure-finding (FF).   In large complex systems and facilities, there are almost always several equipment items—or possibly a whole subsystem or system—that could experience failure and, in the *normal* course of operation, no one would know that such failure has occurred. We call this situation a *hidden failure.* Backup systems, emergency systems, and infrequently used equipments constitute the major source of potential hidden failures. Clearly, hidden failures are an undesirable situation since they may lead to operational surprises and could then possibly initiate an accident scenario via human error responses. For example, an operator may go to activate a backup system or some dormant function only to find that it is not available and, in the pressure of the moment, fail to take the correct follow-up procedure. So, if we can, we find it most beneficial to exercise a prescheduled option to check and see if all is in proper working order. We call such an option a *failure-finding* (FF) task.

Let's look at a couple of examples to illustrate our point about hidden failures and the FF task. First, look at a simple example—the spare tire in our automobile. If you are like me, you don't really worry about a flat spare tire because you have AAA coverage, and are never more than 10–15 minutes away from an ability to get emergency road service—except for that once-a-year trip with the family into "uncharted lands" (e.g., Death Valley!). Again, if you are like me, you *do* check the spare tire before you leave—and that is a failure finding (FF) task. Notice that the only intent in such an action is to determine if the spare tire is in working order or not. We are doing nothing to prevent or retard a flat tire (a TD task) or to measure its incipient failure condition (a CD task). It is or is not in working order. And, if it is not in working order, we fix it. That is the essence of what a failure finding task is all about. (Is it OK? If not, fix it.) In this simple example of the spare tire, notice that there are both TD and CD task alternatives that are available to us. As a TD task, we could elect to check the spare tire pressure on preset intervals which we know to be the limits on the tire's capability to hold the required pressure. (Of course, we really do not know this limit, but we could guess at it anyway in an ultraconservative manner—don't laugh, most maintenance engineers make equally wild guesses every day about such failure states.) Or we could run a pressure-sensing line from the tire to a gauge on the dashboard and closely monitor just when the tire pressure goes below acceptable limits—i.e., a CD task. Why don't we do one of these tasks? In the case of the TD task, it's really too much trouble given the alternative of the AAA service or the FF task on that once-a-year trip. In the case of the CD task, we don't want to pay for that option, and that is why it is never offered by any current U.S. automobile manufacturer. In other words, convenience and/or cost considerations often drive us to use the FF task in lieu of TD or CD tasks in situations where hidden failures could occur.

A more complex example might involve a standby diesel generator that would be called into service if a grid blackout occurred. One complicating factor here is that we cannot pinpoint when the demand will occur. Thus, we usually go into some form of periodic surveillance task where we start the diesel generator set and bring it to a serviceable power condition to maintain a high confidence in its readiness state. Does this absolutely guarantee that it will successfully perform when a real demand occurs? Not necessarily; however, studies have been conducted which show that the probability of successful performance upon demand can be optimized with the selection of the proper interval for a surveillance (failure finding) task.

A third example could involve the use of a particular valve during a power plant startup. This is a rather common situation; a few valves are opened only for flow alignment during startup and are then returned to the closed position until the next plant startup occurs. In this case, we really don't have to spend money to maintain these valves. Rather, we could opt to specify an FF task, and simply assure ourselves that the valves are in working order a few days before they will be used, or fix them if a problem is discovered.

Run-to-failure (RTF). As the name implies, we make a deliberate decision to allow an equipment to operate until it fails—and no preventive maintenance of any kind is ever performed. In other words, the preventive maintenance program has deliberately opted to use a corrective maintenance task as its preferred strategy. There are some limited cases where such a strategy makes common sense, and the details of this strategy will be more fully developed in Chap. 5. Suffice it to say at this juncture that there are three reasons why such a decision can occasionally be made:

1. We can find no PM task that will do any good irrespective of how much money we might be able to spend.

2. The potential PM task that is available is too expensive. It is less costly to fix it when it fails, and there is no safety impact at issue in the RTF decision.

3. The equipment failure, should it occur, is too low on the priority list to warrant attention within the allocated PM budget.

The specifics of how we go about deciding which type of PM task to employ, and examples to illustrate their usage, will be discussed further in Chaps. 5, 6, and 7.

2.4 Preventive Maintenance Program Development

Creating a new PM program, or upgrading an existing PM program, involves essentially the same process. We need to (1) determine what we would ideally like to do in the PM program and (2) take the necessary steps to build that ideal program into our particular infrastructure and put it into action. This process is illustrated in Fig. 2.1, and will be subsequently explained in more detail.

Before anything can happen, we must somehow decide what it is that, ideally, we would like to have in place (i.e., the left side of Fig. 2.1). That is, we should develop what we believe is the optimum PM program without imposing any restrictions that might otherwise lead us to select "second best" choices among alternatives. For example, we should not limit our selections to PM tasks that only fit the skills of the current maintenance technicians. Later, if greater skills are needed, we may decide that the burden of training or the necessity for hiring more skilled personnel is not an option that is available. But, in the ini-

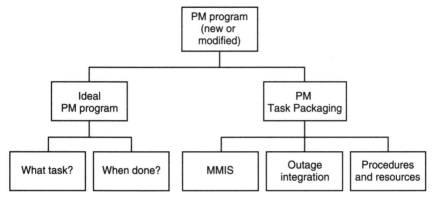

Figure 2.1   Preventive maintenance program development or upgrade.

tial formulation of our PM program, we should go for the best possible (ideal) program that can be conceived so that we have clearly displayed the information that management will need to make the required choices (such as to commit to a training program).

There are only two pieces of information that are required to define the ideal PM program. Specifically, we must identify (1) *what* PM tasks are to be done and (2) *when* each task should be done. Whatever method may be employed to determine "what tasks," it will result in the definition of a series of tasks that are composed of the task types described in Sec. 2.3. We will briefly discuss some of the "what" methods historically employed throughout industry in Sec. 2.5, and ultimately will recommend the RCM approach in Chap. 4—and for a variety of good reasons. Likewise, methods to determine "when done" information will be covered in Sec. 2.5, and some more specific thoughts on the whole issue of task periodicity will be discussed in Chap. 8, Sec. 8.3.

Let us assume for now that we have invoked the recommendations and methodology that are embodied in Chaps. 4–8, and we have defined our ideal PM program. The next job is to incorporate that program, to the maximum extent possible, into the existing corporate infrastructure and assure that it is, in fact, implemented in everyday operations (i.e., integrate the ideal program into the real world, as shown on the right side of Fig. 2.1). In order to do this, there are a series of questions and issues that must be resolved before any implementation can occur. Typically, such issues might include the following:

- Are new procedures, or modifications to existing procedures, required?

- Are all of the standard materials available (tools, lubricants, etc.)?

- Is any special tooling or instrumentation required?

- Are any capital investments required?
- Do we have enough (or too many) people to conduct the program?
- Are the needed skills available? Must training courses be conducted? Is hiring of new skills necessary?
- Does the new/upgraded program affect the spares on-hand (too few, too many)?
- How long will it take to incorporate the new/upgraded program into our MMIS (maintenance management information system)? Is our existing MMIS capable of accepting everything in the new/upgraded program (e.g., tracking time sequenced data in CD tasks)?
- If we must conduct periodically planned full outages, do the tasks and task intervals lend themselves to such a schedule?
- Do new tasks require a periodicity that is a common denominator with other existing task intervals?

Every situation will have its own unique set of questions which may, or may not, look like the ones in this list. But whatever they may be, it is necessary that we carefully accomplish what is called *Task Packaging*—that is, the specific process for integrating the ideal PM task selections into the existing (or modified) corporate infrastructure for the purpose of putting as many of the ideal PM tasks as possible into the daily operating routine.

Only when the ideal PM program is properly integrated with Task Packaging will we have the operational PM program that we had set about to deploy.

2.5   Current PM Development Practices and Myths*

If you were to conduct a survey among the maintenance managers and supervisors in U.S. industry today to ascertain how their PM program came to be what it is, you are likely to accumulate the greatest hodge-podge of ad hoc data and reasoning that has ever been assembled in the world of technical literature! Experience has shown that this statement is not an exaggeration—surprising as it may seem to the uninitiated.

With a few notable exceptions (one being commercial aviation), it is a fairly safe bet to say that the vast majority of existing PM programs cannot be traced to their origins; if they can, the origins thus identified still leave open the fundamental question of just why the PM

---

* *Myth:* A notion based more on tradition or convenience than on fact. (*The American Heritage Dictionary.*)

tasks are being done. (Recall that in Sec. 2.1 we noted that some PM programs are, in fact, reactive programs that are more corrective than preventive in nature. These reactive programs, of course, have a particularly difficult time trying to tell you about their origins.) The implication here is that current PM programs are probably wasting resources doing unnecessary tasks or, conversely, are failing to perform necessary tasks, or perhaps are doing some tasks in a very inefficient fashion (e.g., too frequently). The facts tend to support this implication. The various RCM programs that have been conducted repeatedly show hard evidence to the effect that most PM programs meet all of these implications to one degree or another. We will further discuss some of the experience that has been observed in the following paragraphs.

Failure prevention.   There is still a widespread feeling in the maintenance community that *all* failures can be prevented. This feeling often motivates the use of overhaul tasks without any fundamental questioning or understanding about the failure mechanisms involved. As we will see subsequently, and also in Chaps. 3 and 4, this overextended use of overhauls can be not only unproductive, but even counterproductive (i.e., create failures that were not present before the overhaul). Some things do wear out and/or age deteriorate—but probably not to the extent often perceived. We need to identify the failure mechanisms that are involved and, if wearout or aging mechanisms are absent, we should not try to prevent that which is not present initially—it's a waste of money. As our knowledge base becomes more complete, our ability to prevent failures via PM actions will increase. But we need to carefully examine our ability to prevent failure via PM actions—before we commit resources that could be misplaced.

Experience.   The most common answer given to justify a PM task generally runs like this: "It's been done for 15 years, so it must be good." But did you ever test that hypothesis? Well, ah—no! Let's not mistake for a moment the significant value that resides in experience. But the trick is to use that experience within some logical framework of analysis to lead you to the proper action. The use of "raw" experience is frequently misleading, and perhaps outright wrong.

Judgment.   This is the first cousin of experience. It's the extrapolation of experience to a new or (maybe) related area of equipment. If misplaced experience can lead you into trouble, think what misplaced judgment might do! Judgment usually comes in the statement, "I think this might be a good thing to do." Rarely said, but almost always implied is, "But I'm not sure I can justify why."

Recommendation. This usually comes from the OEM. "The vendor says we ought to do this." The problem with this is that the vendor's recommendations are mainly based on experience and judgment (see previous) and, furthermore, the vendor frequently does not know or understand the specifics of how you will use the equipment. For example, the equipment was designed for steady-state operation, but your application is highly cyclical! Even if the vendor-recommended tasks are correct, they are usually quite conservative on periodicity—especially overhaul intervals. This, of course, might be a good protective measure from *their* point of view.

Brute force. There seems to be a strong feeling in many quarters that if it is physically possible to do something that appears to have a PM characteristic, then it must be a good thing to do. This is "the more the better" syndrome. It can take some weird forms: overlubrication, cleaning when it shouldn't even be touched, part replacement when there is absolutely nothing wrong with the installed part, etc.

Regulation. This is a very difficult area to handle. Most products and services today come under some form of regulatory cognizance—OSHA, EPA, NRC, local PUCs, etc. In their well-meaning ways, these regulators can mandate PM actions that are potentially counterproductive to their objectives. Economics aside for the moment, the major difficulty resides in a lack of appreciation of the risk involved in PM actions (see following). By requiring an owner/operator to do certain tasks, the regulators can actually *increase* the chance for an event (spill, release, etc.) to occur, rather than to help avoid the event. After we educate ourselves, we then need to educate the regulators. Since they are probably here to stay, we should not forget this obligation!

Risk. In this case, the best was saved until the last. There is some conclusive evidence developing that substantiates the "gut feel" of many maintenance engineers that preventive maintenance is, in fact, a potentially risky business. The risk here refers to the potential for creating various types of defects while the PM task is being performed. These defects, or errors, that eventually lead to equipment failures, stem primarily from human errors that are committed during the course of PM task achievement. The risks come in many shapes, sizes, and colors. Typically, they may include:

- Damage to an adjacent equipment during a PM task
- Damage to the equipment receiving the PM task:

    damage during an intrusion for inspection, repair, or adjustment

    installation of a replacement part or material that is defective

misinstallation of a replacement part or material
incorrect reassembly

- Infant mortality of replaced parts or materials

- Damage due to an error in reinstallation of an equipment into its original system

You might wish to reexamine your own records; they are probably replete with evidence of the preceding. And what is especially lethal about this type of "generated" defect is that it usually goes unrecognized—until it causes a forced outage. There has been some published data that illustrates this point—see Fig. 2.2 and Ref. 1. This data involved fossil power plants, and examined the frequency and duration of forced outages *after* a planned or maintenance outage (recall our discussion of MOs in Sec. 2.1). Your attention is drawn to the "Total" column where 56 percent of the forced outages (1772/3146) occurred within one week or less after a planned or maintenance outage! Although these statistics do not reveal exactly how many of those forced outages were due to errors committed during the planned outages, there is strong evidence to conclude that the vast majority were *directly* due to errors during the planned outage. Examples include fan blade balance weights knocked off during cleaning (why were they being cleaned?), improper seal seating on overhauled pumps, and missing parts during a reassembly operation. The message is clear—risk is an inherent factor in PM. One should not perform a PM task unless he or she is really convinced that there is a justifiable reason for doing it, and then there should be attention to assuring that it is done properly.

If we step back from all of the preceding items, and try to summarize their messages, we could conclude the following: All of them are driven by the principle of "What can be done," rather than by the prin-

Duration Time	<1 week	1 to 2 weeks	2 to 4 weeks	>1 month	Total
<1 week	1,705	35	16	16	1,772
1 to 2 weeks	358	5	5	2	370
2 to 3 weeks	258	8	0	1	267
3 to 4 weeks	176	0	0	1	177
1 to 2 months	324	12	2	2	340
2 to 3 months	137	3	0	1	141
>3 months	73	3	0	3	79
Total	3,031	66	23	26	3,146

Figure 2.2 Time between planned or maintenance outage and forced outage versus duration of forced outage.

ciple of "Why should it be done?" The latter is a very key issue and, as we shall see in Chap. 4, was at the source of thinking that ultimately led to RCM.

2.6 PM Program Elements

Figure 2.1 gave us a rather simplistic picture of how a PM program is developed. Let's go one step further and look at some of the supporting management and technical disciplines that are involved in the "Ideal PM Program" and "PM Task Packaging." Please note that the disciplines described in this subsection are generally applicable to any PM program, and are not peculiar to any RCM-driven PM program (although they surely do support the RCM concept that will be developed in later chapters).

Ideal PM program.   There are a host of supporting technologies that could be listed here. Highlighted are those which we believe are most important. They are shown in Fig. 2.3, together with a symbolic picture of how they support the "what task" and "when done" blocks of Fig. 2.1. We will discuss each in more detail.

Failure analysis technology.   Consider this important thought. When a design has been completed and then committed to manufacture and use, the designer believes (and hopes) that the product will operate with 100 percent reliability. In other words, when the design was finalized, the designer had already put his or her best available knowledge into the product. If test and operation of the product prove to be totally successful (a virtually nonoccurring situation), then the designer feels a great deal of satisfaction in the demonstration of the expected product performance. But his or her knowledge base about the product has not been extended—only confirmed. However, if failure occurs, a significant learning opportunity is presented to us. In other words, product malfunctions and failures present us with one of the few important times when we can expand our technical knowledge about the various engineering disciplines. That is, *if* we take advantage of it! And therein lies the reason for the importance that we give to the conduct of a comprehensive program for failure reporting, root cause analysis, and corrective action feedback. Without such a program, it is virtually impossible to establish the proper correction to the problem, or to intelligently decide if some form of preventive maintenance action is possible. Let's look at an example to illustrate this point. Consider a motor-operated valve (MOV) that regulates the flow of fluid in a pipe. We have experienced several jamming failures with this valve. We assume (without the benefit of a thorough root cause analysis) that fluid contamination is the

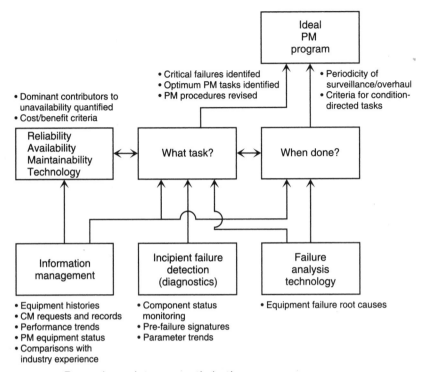

Figure 2.3 Preventive maintenance optimization program.

culprit. So we (1) install a filter upstream of the valve and (2) tighten the requirements on allowable particulates in the fluid reservoir. Result: the valve still jams. This time we get a little smarter and, performing a thorough (microscopic) analysis, discover that the foreign particles involved in the jamming actions are the same material as the valve piston! Upon further examination, we also find that the piston design did not properly chamfer the piston circumference, so the edge was breaking off and the particles were wedging between the piston and cylinder wall. Without this information, we may never have solved the problem, or would have consumed (wasted) valuable resources in a trial and error approach. As shown in App. A, Fig. A.6, a good failure analysis program is also a vital ingredient in the "retain or increase MTBF" portion of an availability improvement program.

Incipient failure detection.   In Sec. 2.3, we discussed the concept of the condition-directed task, and gave several examples of how a CD task might operate. Behind the ability to prudently employ CD tasks is an entire diagnostic technology that is, today, still evolving with new techniques and applications. We believe it is essential to have some form of dedicated effort to follow, understand, and perhaps even contribute to this area that is generally called *predictive maintenance technology.* This subject is, in a sense, a separate book on its own. But to illustrate its content, listed below are several typical tools that constitute elements of predictive maintenance technology:

- Lubricant analysis
- Vibration, pulse, spike energy measurement
- Acoustic leak detection
- Thermal imaging
- Fiber-optic inspection
- Trace element sensing
- Ultrasonic movement sensing
- Debris analysis
- Creep monitoring
- Dynamic radiography measurement
- Stress/strain/torque measurement
- Hyperbaric moisture detection
- Dye penetrant measurement
- Nonintrusive flow measurement
- Microprocessors with expert system software

Information management.   In today's computerized world, it has become necessary to automate the collection, storage, and processing of vital data in order to achieve required levels of operating efficiency. In the operation of large systems, plants, and facilities, such automation is *required* in the conduct and management of the maintenance program. The Maintenance Management Information System (MMIS) is designed to fulfill this need. A typical MMIS will incorporate the following features:

- Automated PM work orders
- PM schedule tracking and measurements
- Corrective maintenance requests and records
- Performance trends
- Failure analysis records
- Condition-directed task measurements and criteria, alerts
- Equipment history

- Industry equipment experience
- Spares/inventory records
- Skill requirements versus skill availability
- PM and CM cost data

Notice that the MMIS is also a key element in PM Task Packaging as shown in Fig. 2.1.

RAM technology. Reliability/Availability/Maintainability (RAM) technology has a broad spectrum of applicability, and can support an availability improvement program (see App. A, Fig. A.6) in many ways. In the area of PM support, RAM models of systems and/or plants can provide the means for predicting and assessing the possible benefits that various PM actions will provide, and evaluating trade-offs that need to be understood in selecting between competing PM options. As a rule, it is *not* suggested that RAM models be developed only for the purpose of PM support, since model development can be a costly task when properly done. Rather, if RAM models have been developed as a part of a broader application and support to an availability improvement program, they should be used for PM support along with their other uses.

PM Task Packaging. This subject will be treated more thoroughly in Chap. 8. Here, we would like to indicate briefly three major elements that must be addressed in Task Packaging.

Task specification. Recall that the output from the ideal PM program in Fig. 2.3 is the "what task" and "when done" information. The task specification is the instrument by which we assure that a complete technical definition and direction is provided to the implementing maintenance organization as to what exactly is required. It is the key transitional document from the ideal to the real world. It is where, for example, we may first learn of certain constraints that will necessitate a departure from the ideal—and will spell out how this must be handled. As a further example, it will detail the data measurement and evaluation requirements for a CD task along with the limiting acceptance criteria, or will specify critical requirements that must be met in a TD overhaul task. In some organizations, the task specification is a very formal written process, complete with documentation change control. In other mature organizations, it is rather informal, and frequently is accomplished in meetings that are a prelude to the second element following.

Procedure. This is the basic document that will guide the field/floor execution of a PM task. In simple PM actions, the procedure may be a one-page instruction, or possibly even a one-line work order authorization. But, in the more complicated PM task, the procedure becomes quite detailed and is considered the "bible" on how the PM task is to be precisely achieved. It should be noted that the risk inherent to PM activities can be controlled and greatly reduced by assuring the development of technically sound and complete task specifications and procedures.

Logistics. Logistics entail a variety of administrative and production support activities. Typical logistic considerations include tooling, spare parts, vendor support, training, documents and drawings, make/buy decisions (i.e., in-house versus contracted work), test equipment, scheduling, regulatory requirements, etc. Clearly, these considerations closely interplay with both the task specification and procedure, and constitute a major portion of what is usually called *maintenance planning.*

In summary, a PM program can be created or upgraded by following the road map of Fig. 2.1. The ideal PM program, supported by key technologies (Fig. 2.3) will produce the "what task" and "when done" information. In Chaps. 4, 5, 6, and 7, we will develop the use of the RCM methodology to supply the "what task" information, and Chap. 8, Sec. 8.3, will discuss the "when done" information. This information must then be subjected to the PM Task Packaging process, described in Chap. 8, to arrive at the specific PM program that can be executed.

The "R" in RCM—Pertinent Reliability Theory and Application

Basic reliability concepts play a key role in the underlying philosophy of RCM, and in its implementation. Not everyone, however, is acquainted with the basic concepts of reliability, especially the use of probability and statistics in formulating key reliability principles. This chapter is intended to introduce (or refresh) the reader on certain of these key principles. In particular, the theory portion of the discussion will be done in simplified, qualitative terms with a more mathematically oriented description presented in App. B for those interested in such detail. In addition to the theory aspect, a comprehensive discussion is included on the use of the failure mode and effects analysis (FMEA) technique since this is employed later in the RCM process.

3.1 Introduction

Reliability-Centered Maintenance (RCM) has been so named to emphasize the role that reliability theory and practice plays in properly focusing (or centering) preventive maintenance activities on the retention of the equipment's inherent design reliability. As the name implies, then, reliability technology is at the very center of the maintenance philosophy and planning process. It thus seems relevant that we discuss some pertinent aspects of the reliability discipline as a prelude to the specific discussions on RCM that are covered in subsequent chapters.

Our objectives in this chapter are to familiarize the reader with what is commonly called *reliability engineering,* and then to describe two specific aspects of the discipline that form the application backbone of the RCM methodology: first, basic reliability theory concepts, and second, one of the key reliability tools known as *failure mode and effects analysis* (or FMEA). The theory portion of the discussion will be done in simplified, qualitative terms, but App. B has been included for those interested in a more mathematically oriented description. References 2 and 3 also contain some excellent material for further insights on the subject.

3.2 Reliability and Probabilistic Concepts

The generally accepted formal definition of reliability is as follows:

> *Reliability* is the probability that a device will satisfactorily perform a specified function for a specified period of time under given operating conditions.

Thus, satisfactory performance occurs under three specified constraints:

1. Function
2. Time
3. Operating conditions (environment, cyclic, steady state, etc.)

Further, achievement of satisfactory performance is a probabilistic notion, and this introduces the concept of a chance element to the reliability discipline. Satisfactory performance is not a deterministic attribute; it is not the case that it either will or will not happen with absolute certainty. Thus, we must deal with the *probability* that a device will succeed or fail under specified constraints since it is impossible for us to state with absolute certainty, before the fact, just which outcome will occur.

Consciously or unconsciously, we all deal with probability on a daily basis. For example:

- My car will most probably start without trouble in the morning.
- There is a good chance of afternoon thunderstorms.
- My chance of being involved in an aircraft crash is very small even though I travel frequently by commercial air.
- A single draw from a deck of cards will probably not be an ace.
- There is some finite chance of another space shuttle accident during the next 100 flights.
- It is highly probable that all of us have made, or heard someone make, one or more of the above statements.

Notice that *none* of the preceding statements can be made with absolute certainty. However, within the mathematical principals of probability and statistics, it is possible to assign values between 0 and 1.0 to all of these statements if we are willing to work at it hard enough (and spend the money to gather data that is pertinent to each situation). In some instances, it is rather easy to do (e.g., probability that a single draw will not be an ace is 48/52). In other instances, it becomes difficult to quantify the probability involved, and sophisticated techniques plus costly testing are required (e.g., probability of a shuttle accident in the next 100 missions). It is indeed rare that we encounter a truly deterministic situation, but some do exist within our current domain of knowledge. For example:

- The sun will rise tomorrow at 6:21 a.m.
- We will die. We will pay taxes.
- Water and oxygen are essential to human survival.

Most of the situations that we encounter in the engineering world have the chance element aspect associated with them. For example, material properties vary, physical environments vary, loads vary, power and signal inputs vary. Some of our basic physical laws can be treated as deterministic—water seeks the lowest accessible level, objects float when they displace their own weight, etc. But when we apply these laws to everyday products, the product performance over time becomes a probabilistic situation. Thus, one would conclude that probability must be a well understood and universally applied discipline in engineering. Unfortunately, not so, although the past decade has made considerable strides in developing and applying probabilistic design concepts, and the reliability discipline has also made some quantum jumps in probabilistic applications. This latter progression has been motivated by a variety of factors, including issues of safety, regulation, warranty, litigation, and the evolving world competitive market.

So just what are some of the basic aspects that must be considered in the probabilistic sense, and how is this done? Recognize, of course, that complete undergraduate and graduate degrees are given in probability and statistics; anything we can say here must be very abbreviated and simplistic. But a sense of how probability operates will be briefly discussed here.

Most probabilistic events of interest involve counting exercises, ranging from simple to complex. The probability of drawing specified cards from a fair deck illustrates this point well:

$$P \text{ (ace in a single draw)} = \frac{\text{number of aces in deck}}{\text{number of cards in deck}} = \frac{4}{52}$$

The problem becomes more complicated if we ask for the probability of drawing a jack *or* a diamond, since the jack of diamonds satisfies both requirements. If we want to know the probability of being dealt a full house in a five-card-draw poker game, added complexities in the counting process are encountered.

Some probability problems involve a question that requires a knowledge of a whole complex population of data. For example, if one should randomly select a person off the street, what is the probability that he or she would be 5 ft 10 in or taller? This kind of question introduces the notion of a population height distribution and the necessity to have some reasonable formulation of that distribution in order to answer such a question. We might develop such a distribution for our particular question by examining military service medical records and plotting the distribution of height data therein with the assumption that such is also valid for the whole U.S. population. Such a plot would most likely look like Fig. 3.1, which has already been converted to a *probability density function* (pdf). This particular pdf is the familiar bell-shaped or Normal distribution which happens to be the distribution that we find for many population characteristics of interest. In Fig. 3.2, we find the answer to our question in the shaded area. Since the area under the entire curve in Fig. 3.1 is equal to 1.0, then one-half of the area in Fig. 3.2 represents our answer, or 0.5. If our question had been directed to the probability that he or she would be between 5 ft 10 in and 6 ft 0 in, then Fig. 3.3 would provide the answer by calculating the area in the cross-hatched section shown. All of this can be done with mathematical precision once the basic pdf for the parameter in question (in this case, height of the U.S. population) is known.

There are a variety of pdf's that can be used to describe the probability of an event of interest. For example:

- The probability of exactly X heads in Y flips of a fair coin will use the Binomial distribution as the basis to describe a population of trials where each trial has only two possible outcomes. Clearly, a single flip

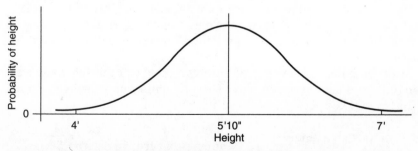

Figure 3.1 The gaussian (normal) pdf for population height.

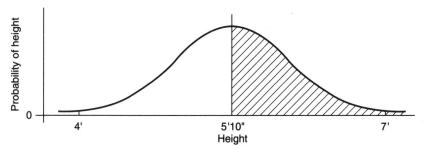

Figure 3.2 Calculating probability (height ≥ 5'10").

of a fair coin is 50/50 for a head or tail. But to calculate the probability of 3 heads in 10 flips of a fair coin before any coin flipping has occurred requires some more sophisticated calculations.

- If we know the average rate at which phone calls come into a switchboard, we can use a Poisson distribution to calculate the probability that the switchboard will receive 0, 1, 2, 3, or N calls during a 60-minute period. Such information is very useful in decisions on staffing levels or training/skill requirements for hiring switchboard operators.

Several different distributions or pdf's exist, and they tend to be used for different kinds of populations and events that can be described. In Sec. 3.4, one specific distribution, the Exponential, will be of particular interest in our discussion of reliability theory.

The point to all of this, again, is that reliability is a probabilistic concept, and thus some basic understanding and appreciation of probability is very much in order.

3.3 Reliability in Practice

It is basic to the notion of reliability that we have some appreciation of just how this probabilistic aspect in the real world might affect the

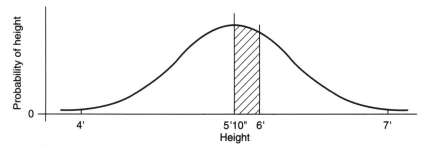

Figure 3.3 Calculating probability (6'0" ≤ height ≥ 5'10").

products that we design, build, and operate. Figure 3.4 is a picture that displays this reality. Suppose we have a product or system composed of N identifiable elements or devices (Fig. 3.4 shows cases for $N = 10$, 50, 100, and 400). None of these elements is "perfect," hence the curve presents cases (on the x-axis) for individual element reliability values

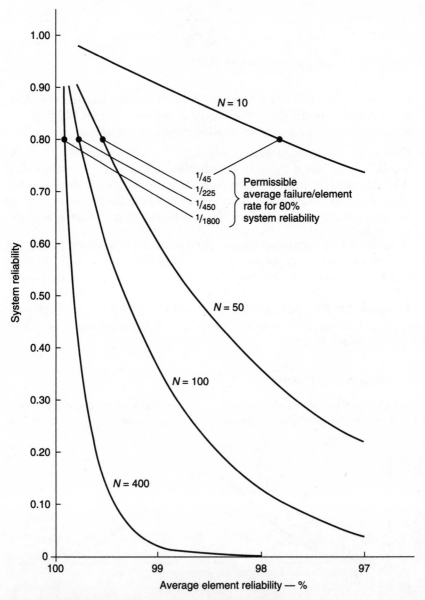

Figure 3.4   Element reliability impact on system reliability.

between 100 and 97% (which, as individual element values, are considered very good in many systems). The system requires that all N elements perform satisfactorily for system success. The probability of system success—i.e., the value of system reliability—is shown on the y-axis. Two things are quite apparent in this display:

1. As system complexity increases (i.e., as N increases), system reliability drops dramatically even for average element reliability values greater than 99 percent. Thus, if a system must contain a large number of individually required elements, these elements must, per se, have very high reliabilities, or we can expect the system to fail frequently.

2. Even if the system is relatively simple (say, $N = 50$), the individual average element reliability needs only to drop slightly and the system reliability drops significantly!

Thus, the need to concentrate on product reliability is not just a PR game—it is very real in terms of ultimately providing the expected customer satisfaction that is emphasized in the TQM discussion in App. A.

As an aside, you might already have noticed that one aspect of the Japanese product strategy is based on the two items just mentioned—that is, (1) keep it simple and (2) have very high individual element reliability.

The important question thus becomes, "How do we achieve high system reliability, and what are the key ingredients that must be addressed?" First and foremost, it must be recognized that reliability is a design attribute. By this, we mean that product reliability is established by how well (or poorly) the design process is accomplished. Reliability cannot be fabricated, tested, or inspected into a product. The design or, more broadly, the product definition (which also encompasses how the product must be operated and maintained) is the sole determinant in setting the inherent, or upper level, of reliability that can be achieved. Fabrication, assembly, test, operation, and even maintenance can only degrade the inherent reliability if they are not performed properly—but none of these activities can enhance it beyond the capability established by the basic design and product definition.

Programs, which are frequently organized under the title of "Reliability Engineering," are often employed to bring together a variety of technical and management functions that will concentrate on guiding and assisting the basic engineering functions in achieving the expected product reliability performance. Some of these reliability engineering functions typically include the following activities:

- Comprehensive review of product specifications to assure that all reliability objectives and supporting requirements are properly included

- In-depth reliability analyses of the design concept and detail via reliability prediction models, design trade studies, failure mode and effects analysis, critical function analysis, design life analysis, proper use of design standards, incorporation of redundancy and other forms of design margin, etc.

- Continuous review and control of potential risk contributors to the product—e.g., adherence to proper part and material applications, establishment of proper manufacturing and quality control processes, structuring meaningful product test programs, control of design changes, and assurance that failures and problems are thoroughly analyzed and fed back to the design for necessary corrective actions

It is during the product design and development phase that preventive maintenance tasks are initially specified. The RCM methodology described in subsequent chapters is a highly effective method for developing these initial PM task specifications. Clearly, the design process should recognize the importance of a proper PM program in retaining the inherent reliability of a product. PM actions ranging from simple lubrication tasks to the more complex replacement of certain life-limited parts are necessary ingredients in the retention of inherent reliability. Unfortunately, this aspect of the design process is often relegated to a secondary priority, and products are then fielded with a less than adequate PM program—and thus, a less than reasonable probability that they will operate up to the customer expectations for reliability. Many products and systems in operation today fall into this category. However, it is still possible to apply the RCM methodology to these products and systems, thereby upgrading their PM programs and ultimately realizing the full potential of the inherent design reliability.

3.4   Some Key Elements of Reliability Theory

In Sec. 3.2, we saw that probability calculations derive from a counting process, and that this process may require a knowledge of population data which will describe the parameter of interest. In reliability, the population that will enable us to calculate reliability values is the failure versus time data. In other words, we need to understand the time distribution for how a large number of devices will fail (or die). If we can accumulate enough data to define or approximate such a distribution, we can define a population density function of the failures—or, in this case, a failure density function (fdf). In the mathematics of reliability, the fdf is usually designated as $f(t)$. Once we know $f(t)$, we can calculate reliability, unreliability, and two very important parameters called the death rate and the mortality rate. These latter two terms derive from actuarial statistics which are employed in the insurance business in order to set policy premium payments.

The *death rate* is defined as the death (or failure) frequency with respect to the *original* population, while the *mortality rate* is the failure frequency with respect to the *surviving* population at some time of interest. A typical life insurance example can illustrate this distinction quite easily. Suppose we look at 1 million people born in 1929, and the records tell us that 10,000 of these people died in 1989. The *death* rate in 1989 for people born in 1929 is:

$$\frac{10,000}{1,000,000} = \frac{1}{100}$$

Now, if only 200,000 of the original 1 million are living on January 1, 1989, then the mortality rate in 1989 for people 60 years old is:

$$\frac{10,000}{200,000} = \frac{1}{20}$$

Obviously, insurance premiums are established on the basis of the mortality rate which is usually labeled as $h(t)$, or just λ, and is commonly referred to as the *instantaneous failure rate*, or just failure rate.

In reliability problems, λ is the parameter of interest to us. In other words, we want to know the probability that devices currently in operation will continue to operate satisfactorily for the next T hours. Or conversely, what is their probability of failure? If we know the fdf or f(t) for the device, we can calculate all of these values.

In reliability, there is one fdf of special interest that is called the *exponential fdf.* This special interest arises for two reasons:

1. There is some substantial evidence that many devices (especially electronics) follow the exponential fdf law.

2. Mathematically, the exponential fdf is the easiest to handle. Because of this feature, we often assume that some product, system, or device follows the exponential law—only to later find that such is not true, and we have thus miscalculated the product reliability.

The specific feature that makes the exponential fdf so easy to handle is that the mortality or failure rate, λ, is a *constant over time* (rather than varying with time). This mathematical nicety means that, in the hardware world, device or product failure is a *random* process which occurs, on average over an extended period, at some fixed time interval. Stated differently, if λ is a constant, then the failures are independent of time, and will neither increase nor decrease in frequency as the product or device population ages. This feature has some profound implications on preventive maintenance, and these are discussed in Chap. 4, Sec. 4.2. Further, the reciprocal of λ, $1/\lambda$, is the mean of the exponential fdf, and is called the *mean time between failure* or MTBF. All pdfs (or fdfs) have a mean value called the *mean time to failure* or MTTF. With the exponential fdf, MTTF = MTBF, but only with the exponential is the MTBF a constant over time. With other fdfs, MTTF is a single value that occurs only once for the distribution represented. Thus, if we do *not* have an exponential law governing the failure history for a device or product, we will experience different failure rates depending upon where we are in the device or product life cycle. In these instances, it is important to know the details of the failure mechanisms and their causes so that proper design, maintenance, or operation actions can be taken to achieve the specified reliability. There is a generally accepted concept in reliability that attempts to put both the constant and nonconstant λs together to describe a typical device or product life cycle. It is called the *bathtub curve* (see Fig. 3.5). The name clearly derives from its shape which evolves from the following three scenarios:

1. In the early stages of product deployment, there is some residual of substandard parts, materials, processes, and workmanship that escapes the factory test and checkout actions, and thus remains in the product at its point of initial use and operation. These substandard items generally surface rather quickly relative to the total product lifetime, but initially they produce a failure rate that is larger than the expected long-term failure rate. As these problems

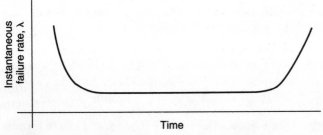

Figure 3.5 The reliability "bathtub curve."

surface and are removed, the population failure rate will decrease and a stabilization of the population λ will occur. This first phase of the cycle is called the *infant mortality stage.*

2. When the population stabilization is complete, the constant failure rate phase described previously takes over. Product failures are now random in nature, and we have stabilized at the level of inherent reliability of the product. The product population, *on average,* has a constant MTBF, but because of the randomness in failure occurrence, we can predict neither the precise time nor the exact nature of the failures that will ultimately occur.

3. As the product operating life progresses, several potential failure mechanisms may develop which are no longer random in nature. In fact, they are very time- or cycle-dependent, and lead to product aging and wearout. These mechanisms include items such as material wear, fatigue deterioration, grain structure changes, and material property changes. When this happens, the population failure rate will again start to increase, and we see that the product may be nearing its end of useful life if the repair or replacement of the affected parts or devices is extensive and costly. This third phase of the scenario is called the *aging and wearout stage.*

Whether or not an item of equipment has the infant mortality and/or aging and wearout stages likewise has a profound impact on preventive maintenance strategy, and this is discussed in Chap. 4, Sec. 4.2. Often, the case may be that infant mortality and/or aging and wearout are known, with a reasonable engineering confidence, to exist. But the times at which these stages occur in the product lifetime are not well defined. In this situation, we also have some decisions to make regarding the choice of a PM task and its periodicity. More is said about this in later discussions including the discussion on age exploration in Chap. 8, Sec. 8.3.

In summary, the key elements of reliability theory that are germane to the RCM methodology are as follows:

1. Knowledge of a product or device fdf allows the calculation of the reliability parameters that may be of interest.

2. A key parameter in this regard is the failure rate, λ.

3. One specific fdf frequently quoted and employed is the exponential fdf wherein λ = constant, and is therefore independent of time. (Often the exponential is assumed when, in reality, it is not the proper fdf.)

4. There is a generally accepted depiction of a typical product life cycle known as the *Bathtub Curve.*

5. Whether a product or device follows the Bathtub Curve can have a profound impact on the proper selection of PM tasks.

3.5   Failure Mode and Effects Analysis (FMEA)

The FMEA is generally recognized as the most fundamental tool employed in reliability engineering. Because of its practical, qualitative approach, it is also the most widely understood and applied form of reliability analysis that we encounter throughout industry. Additionally, the FMEA forms the headwaters for virtually all subsequent reliability analyses and assessments because it forces an organization to systematically evaluate equipment and system weaknesses, and their interrelationships that can lead to product unreliability.

The FMEA embodies a process that is intended to identify equipment failure modes, their causes, and finally the effects that might result should these failure modes occur during product operation.* Traditionally, the FMEA is thought of as a design tool whereby it is used extensively to assure a recognition and understanding of the weaknesses (i.e., failure modes) that are inherent to a given design in both its concept and detailed formulation. Armed with such information, design and management personnel are better prepared to determine what, if anything, could and should be done to avoid or mitigate the failure modes. This information also provides the basic input to a well-structured reliability model that can be used to predict and measure product reliability performance against specified targets and requirements.

The delineation of PM tasks is also based on a knowledge of equipment failure modes and their causes. It is at this level of definition that we must identify the proper PM actions that can prevent, mitigate or detect onset of a failure condition. Specifying PM tasks without a good understanding of failure mode and cause information is, at best, nothing more than a guessing game. Hence, the FMEA will play a vital role in the RCM process, and this will be developed in more detail in Chap. 5.

---

* By definition, a *failure mode* is a simple two or three word descriptor of *what* went wrong; the failure cause is a further descriptor of *why* the failure mode occurred in the first place; the failure effect is *how* the failure mode impacts both the device involved as well as other equipments and systems that may be associated with it.

How do we perform the FMEA? First, it should be clear by now that a fairly good understanding of the equipment design and operation is an essential starting point. The FMEA process, itself, then proceeds in an orderly fashion to qualitatively consider the ways in which the individual parts or assemblies in the equipment can fail. These are the failure modes that we wish to list, and are physical states in which the equipment could be found. For example, a switch can be in a state where it cannot open or close. The failure modes thus describe desired functions of the device which have been lost. Alternately, when sufficient knowledge or detail is available, failure modes may be described in more specific terminology—such as "jammed," or "actuating spring broken." Clearly, the more precise the failure mode description, the more understanding we have for deciding how it may be eliminated, mitigated, or accommodated. Although it may be difficult to accurately assess, we also attempt to define a credible failure cause for every failure mode (maybe more than one if deemed appropriate to do so). For example the failure mode "jammed" could be caused by contamination (dirt), and the "broken spring" could be the result of a material-load incompatibility (a poor design).

Each failure mode is then evaluated for its effect. This is usually done by considering not only its local effect on the device directly involved, but also its effect at the next higher level of assembly (say, subsystem) and, finally, at the top level of assembly or product level (say, system or plant). It is usually most convenient to define two or three levels of assembly at which the failure effect will be evaluated in order to gain a full understanding of just how significant the failure mode might be if it should occur. In this way, the analyst gains a bottoms-up view of what devices and failure modes are important to the functional objectives of the overall system or product. A typical FMEA format is shown on Fig. 3.6.

By way of example, an FMEA is shown on Fig. 3.7 which is based on the simple lighting circuit schematic shown in Fig. 3.8. In this instance, the FMEA is conducted at the system level due to its simplicity, and we just move around the system circuit, device by device. In a more complex analysis, we might devote an entire FMEA to just one device, and break it into its major parts and assemblies for analysis. A pump or transformer are examples of where this might be done.

Frequently, FMEAs are extended to include other information for each failure mode—especially when the FMEA is conducted in support of a design effort. These additional items of information could include:

- failure symptoms
- failure detection and isolation steps
- failure mechanisms data (i.e., microscopic data on the failure mode)

| EQUIPMENT | FAILURE MODE | FAILURE CAUSE | FAILURE EFFECTS | | |
I.D. #	DESCRIPTION			LOCAL	SYSTEM	UNIT

Figure 3.6 Failure mode and effects analysis.

Component	Mode	Effect	Comment
1. Switch A1	1.1 Fails open 1.2 Fails closed	1.1 System fails 1.2 None	1.1 Cannot turn on light. 1.2 If A2 also fails closed, then system fails by premature battery depletion.
2. Switch A2	(same as A1)	(same as A1)	(same as A1)
3. Light Bulb C	3.1 Open filament 3.2 Shorted base	3.1 System fails 3.2 System fails; possible fire hazard	3.1 Cannot turn on light. 3.2 Cannot turn on light. May cause secondary damage to rest of system.
4. Battery B	4.1 Low charge 4.2 No charge 4.3 Over-voltage charge	4.1 System degraded; dim light bulb 4.2 System fails 4.3 System fails by secondary damage to Light Bulb C	4.1 May be precursor to "no charge." 4.2 Cannot turn on light. 4.3 Secondary damage to Light Bulb C caused by over-current.

Figure 3.7 Simple FMEA.

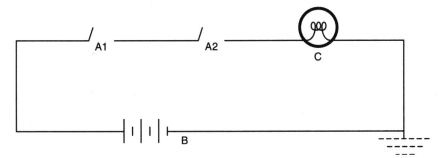

Figure 3.8 Simple circuit schematic.

- failure rate data on the failure mode (not always available with the required accuracy)
- recommended corrective/mitigation actions

When a well-executed FMEA is accomplished, a wealth of useful information is generated to assist in achieving the expected product reliability.

RCM—A Proven Approach

In this chapter, we will introduce the basic concepts that constitute what is known as *Reliability-Centered Maintenance*. Initially, however, we will briefly discuss how PM has evolved in the industrial world, and most importantly, we will look at how one of the basic tenets of reliability engineering—the "bathtub curve"—can and should influence the formulation of PM tasks. Next, we will look at how the commercial aviation industry was historically the motivating force behind the creation of the RCM methodology during the Type Certification process for the 747 aircraft in the 1960s. Finally, we will itemize the four basic features that constitute the necessary and sufficient conditions or principles that define RCM, and discuss some of the cost-benefit considerations that can accrue through the use of RCM.

4.1 Some Historical Background

If we look back to the days of the Industrial Revolution, we find that the designers of the new industrial equipment were also the builders and operators of that equipment. At the very least, they had a close relationship with the hardware that evolved from their creative genius, and as a result they truly did "know" their equipment—what worked, how well, and for how long; what broke, how to fix it, and, yes, how to take certain reasonable (not too expensive) actions to prevent it from breaking. In the beginning, then, experience did in fact play the major role in formulating PM actions. And, most importantly, these experience-based actions derived from those people who had not just maintenance experience, but also design, fabrication, and operation knowledge. Within the limits of then available technology, these engineers were usually correct in their PM decisions.

As industry and technology became more sophisticated, corporations organized for greater efficiency and productivity. This, of course, was necessary and led to numerous advantages that ultimately gave us the high-volume production capability that swept us into the twentieth century. But some disadvantages occurred also. One of these was the separation of the design, build, and operate roles into distinct organizational entities where virtually no one individual would have the luxury of personally experiencing the entire gamut of a product cycle. Thus, the derivation of PM actions from experience began to lose some of its expertise.

Not to worry! Another technology came along to help us—reliability engineering. The early roots of reliability engineering trace back to the 1940s and 1950s. Much of its origin resides in the early work with electronic populations where it was found that early failures (or infant mortalities) occurred for some period of time at a high but decreasing rate until the population would settle into a long period of constant failure rate. It was also observed that some devices (e.g., tubes) would finally reach some point in their operating life where the failure rate would again sharply increase, and aging or wearout mechanisms would start to quickly kill off the surviving population. (This scenario, of course, also very accurately describes age-reliability characteristics of the human population.) Engineers, especially in the nonelectronic world, were quick to pick up on this finding, and to use it as a basis for developing a maintenance strategy. The picture we have just described is the well-known *bathtub curve*. Its characteristic shape (seen previously in Chap. 3, Fig. 3.5) led the maintenance engineer to conclude that the vast majority of the PM actions should be directed to *overhauls* where the equipment would be restored to like-new condition before it progressed too far into the wearout regime.

Thus, until the early 1960s, we saw equipment preventive maintenance based in large measure on the concept that the equipment followed the bathtub shape, and that overhaul at some point near the initiation of the increasing failure-rate region was the right thing to do.

Some additional historical perspective on the evolution of reliability engineering can be found in Refs. 4 and 5.

4.2   The "Bathtub Curve" Fallacy

As this title suggests, all may not be totally well with the bathtub curve. True, some devices *may* follow its general shape, but the fact is that more has been assumed along those lines than has actually been measured and proven to be the case. As those with even a cursory knowledge of statistics and reliability theory can attest, this is not surprising, because large sample sizes are required in order to accurately develop the population age-reliability characteristics of any given device, component, or system. And such large samples, with recorded data on operating times and failures, are hard to come by.

The commercial aviation industry, however, does have fairly large populations of identical or similar components in their aircraft fleets— components that are common to several aircraft types. And, as an industry, they have made some deliberate and successful efforts to accumulate a database of operating history on these components. Such a database is driven by several factors, not the least of which are safety and logistics considerations. As a part of the extensive investigation that was conducted in the late 1960s as a prelude to the RCM methodology, United Airlines used this database to develop the age-reliability patterns for the nonstructural components in their fleet. This was done as a part of the more general questioning that preceded RCM concerning whether airline equipments did, in fact, follow the bathtub curve. Specifically, failure density distributions were developed from the component operating history files, and the hazard rate (or instantaneous failure rate) was derived as a function of time. The results of this analysis are summarized in Fig. 4.1 (from Ref. 6).

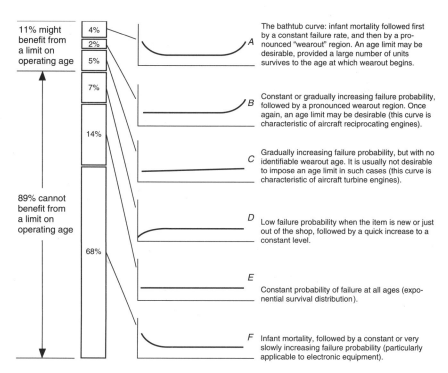

Figure 4.1   Age–reliability patterns for nonstructural aircraft equipment (United Airlines).

These results came as a surprise to almost everyone—and continue to do so even today when people see these results for the first time. The significance of these results, and their potential importance to the maintenance engineer, cannot be stated too strongly. Let's examine these more closely, assuming for the moment that these curves may be characteristic of your plant or system:

1. Only a very small fraction of the components (4 percent) actually replicated the traditional bathtub curve concept (curve A).

2. More significantly, only 6 percent of the components experienced a distinct aging region during the useful life of the aircraft fleet (curves A and B). If we are generous in our interpretation, and allow that curve C also is an aging pattern, this still means that only 11 percent of the components experienced an aging characteristic!

3. Conversely, 89 percent of the components never saw any aging or wearout mechanisms developing over the useful life of the airplanes (curves D, E, and F). Thus, while common perceptions tend toward the belief that 9 of 10 components have "bathtub" behavior, the analysis indicated that this was completely reversed when the facts were known.

4. Notice that 72 percent of the components, however, did experience the infant mortality phenomenon (curves A and F).

5. And the most common grouping, 68 percent, was starting to look like the bathtub, but never got to the aging region (curve F).

What does all of this mean? Quite a bit! First, recall that a constant failure rate region (curves A, B, D, E, and F all have this region) means that the equipment failures in this region are random in nature—that is, the state of the art is not developed to the point where we can predict what failure mechanisms may be involved, nor do we know precisely when they will occur. We only know that, on average in a large population, the hazard rate (or the mean time between failure) is a constant value. Of course, we hope that this constant-failure-rate value is very small, and we thus have a very reliable set of components in our system. But, for the maintenance engineer, these constant-failure-rate regions mean that overhaul actions will essentially (short of luck) do very little, if anything, to restore the equipment to a like-new condition. In this constant-value region, overhaul is usually a waste of money because we really do not know what to restore, nor do we really know the proper time to initiate an overhaul. (In the constant-failure-rate region, any time you might select is essentially the wrong time!) Second, and worse yet, is that these overhaul actions will actually be harmful because, in our haste to restore the equipment to new, pristine conditions, we have inadvertently pushed it back into the infant-mortality region of the curve. (See Fig. 2.2 in Chap. 2.) In this specific study, for example, overhaul actions on 72 percent of the components (curves A and F) would be susceptible to this counterproductive situation. A third point relates to the periodicity that should be specified for an overhaul task when such an action is considered to be the correct step to take. For example, if a component is either a curve A or B type, we want to assure that the overhaul action is not taken too soon—or again, we may be wasting our resources. Often, we do not know what the correct interval should be, or even if an overhaul PM task is the right thing to do. Why? Because we do not have sufficient data to tie down the age-reliability patterns for our equipment. In these instances, we may wish to initiate an Age Exploration program, and more on this topic will be covered in Chap. 8, Sec. 8.3.

In summary, we should be very careful about selecting overhaul PM tasks because our equipment may not have an age-reliability pattern that justifies such tasks. In addition, overhauls are likely to cause more problems than they prevent if aging regions are not present. When data is absent to guide us on this very fundamental and important issue, we should initiate an Age Exploration program and/or the collection of data for statistical analyses that will permit us to make the right decisions. It is indeed a curious (and unfortunate) fact that in today's world of modern technology, one of the least understood phenomenon about our marvelous machines is how and why they fail!

4.3   The Birth of RCM

RCM epitomizes the old adage that "necessity is the mother of invention."

In the late 1960s, we found ourselves on the threshold of the jumbo jet aircraft era. The 747 was no longer a dream; the reality was taking shape as hardware at the Boeing factory in Seattle. The licensing of an aircraft type (called *Type Certification* by the FAA) requires, among its many elements, that an FAA-approved preventive maintenance program be specified for use by all owners/operators of the aircraft. No aircraft can be sold without this Type Certification by the FAA. The recognized size of the 747 (three times as many passengers as the 707 or DC-8), its new engines (the large, high bypass ratio fan jet), and its many technology advances in structures, avionics, and the like, all led the FAA to initially take the position that preventive maintenance on the 747 would be very extensive—so extensive, in fact, that the airlines could not likely operate this airplane in a profitable fashion. This development led the commercial aircraft industry to essentially undertake a complete reevaluation of preventive maintenance strategy. This effort was led by United Airlines who, throughout the 1960s, had spear-headed a complete review of why maintenance was done, and how it should best be accomplished. Names like Bill Mentzer, Tom Matteson, Stan Nowland, and Harold Heap, all of United Airlines, stand out as the pioneers of this effort (Refs. 6, 7, 8). What resulted from this effort was not only the thinking derived from the curves in Fig. 4.1, but also a whole new approach that employed a decision-tree process for ranking PM tasks that were necessary to preserve critical aircraft functions during flight. This new technique for structuring PM programs was defined in MSG-1 (Maintenance Steering Group-1) for the 747, and was subsequently approved by the FAA. The MSG-1 was able to sort out the wheat from the chaff in a very rational and logical manner. When this was done, it was clear that the economics of preventive maintenance on a 747-sized aircraft were quite viable—and the 747 became a reality.

The MSG-1 was so successful that its principles were applied in MSG-2 to the Type Certification of the DC-10 and L-1011. In recent times, MSG-3 has developed the PM program for the 757 and 767. Versions of the MSG format have likewise guided the PM programs for the Concorde, Airbus, 737-300/400/500, and various retrofits to aircraft such as the 727-200, DC-8 stretch, and DC-9 series.

In 1972, these ideas were first applied by United Airlines under Department of Defense (DOD) contract to the Navy P-3 and S-3 aircraft and, in 1974, to the Air Force F-4J. In 1975, DOD directed that the MSG concept be labeled "Reliability-Centered Maintenance," and that it be applied to all major military systems. In 1978, United Airlines produced the initial RCM "bible" (Ref. 6) under DOD contract. More recent discussions on RCM applications in commercial aviation are found in Refs. 9 and 10.

Since then, all military services have employed RCM on their major weapons systems. RCM specifications have been developed (e.g., Ref. 11), the Air Force Institute of Technology (AFIT) offers a course in RCM, and the Navy has published an RCM handbook (Ref. 12).

The most recent use of RCM has occurred in the utility (electric power generation) industry. In 1983, the Electric Power Research Institute (EPRI) initiated RCM pilot studies on nuclear power plants (Refs. 13, 14, 15). Since these early pilot studies, several full-scale RCM applications have been initiated in commercial nuclear and fossil power plants (e.g., Refs. 16, 17, 18, 19, 20).

Clearly, the development of RCM has been an evolutionary process, and some 30 years have passed during which RCM has become a mature process in commercial aviation and DOD, and an embryo process in power generation plants. Basically, no other industry has yet fully embraced RCM, in spite of its proven track record. Hopefully, this book will help to change that picture.

4.4   What Is RCM?

In Chap. 2, Sec. 2.5, we briefly examined some of the more prominent practices and myths that currently constitute the basis for PM program development. We summarized by saying that these practices and myths are driven, in large measure, by the overriding consideration and principle of "what can be done?"—but rarely by traceable decisions such as "why should it be done?" Stated another way, we could say that the overriding motivation of current PM practices is to "preserve equipment operation." Until recently, this has resulted in little, if any, consideration as to why we take certain PM actions and what, if any, priority should be assigned to the expenditure of PM resources. Almost without fail, maintenance planning starts directly with the equipments and seeks to specify as quickly as possible the various things that are felt necessary to "keep it running" (sometimes without a conscious evaluation of the function that is served or consideration of a cost-benefit decision).

On the other hand, RCM is not just another way to do this same old thing all over again. It is very different in some fundamental aspects from what is today the norm among maintenance practitioners, and requires that some very basic changes in our mind-set must occur. As you will see in a moment, the basic RCM concept is really quite simple, and might be characterized as organized engineering common sense.

There are four features that define and characterize RCM, and set it apart from any other PM planning process in use today. Each of these is defined and discussed here.

Feature 1.   The most important of the four RCM features is also perhaps the most difficult to accept because it is, at first glance, contrary to our ingrained notion that PM is performed to preserve equipment operation. *The primary objective of RCM is to preserve system function.* Notice that this objective does not initially deal with preservation of equipment operation. Of course, we will ultimately preserve system function by preserving equipment operation, but not as the initial step in the RCM process. By first addressing system function, we are saying that we want to know what the expected output is supposed to be, and that preserving that output (function) is our primary task at hand. This first feature enables us to systematically decide in later stages of the process just what equipments relate to what functions, and will not assume a priori that "every item of equipment is equally important," a tendency that seems to pervade the current PM planning approach.

Let's look at a couple of simple examples to illustrate the inherent value associated with the "preserve system function" concept. First, compare two separate fluid transfer trains in a process plant where each train has redundant legs. Train A has 100 percent capacity pumps in each leg, and train B has 50 percent capacity pumps in each leg. As the plant manager, I tell you, the maintenance director, that your budget will allow PM tasks on either train A pumps or train B pumps, but not both. What do you do? Clearly, if you don't think function, you are in a dilemma, since your background says that your job is to keep all four pumps up and running. But if you do think function, it is clear that you must devote the defined resources to the train B pumps since loss of a single pump reduces capacity by 50 percent. Conversely, a loss of one pump in train A does not reduce capacity at all, and also in all likelihood allows a sizable grace period to bring the failed pump back to operation. As a second example, let's examine more closely just what function is really performed by a pump. The standard answer is to preserve pressure or maintain flow rate—a correct answer. But there is another, more subtle, function to preserve fluid boundary integrity (a passive function). In some cases, allocation of limited resources to PM tasks for the passive function could be more important than keeping the pump running (e.g., when the fluid is toxic or radioactive). Again, if you don't think function, you may miss drawing the proper attention to the passive boundary integrity function.

Feature 2. Since the primary objective is to preserve system function, then *loss of function or functional failure* is the next item of consideration. Functional failures come in many sizes and shapes, and are not always a simple "we have it or we don't" situation. We must always carefully examine the many in-between states that could exist, because certain of these states may ultimately be very important. The loss of fluid boundary integrity is one example of a functional failure that can illustrate this point. A system loss of fluid can be (1) a very minor leak that may be qualitatively defined as a drip, (2) a fluid loss that can be defined as a design basis leak—that is, any loss beyond a certain GPM value will produce a negative effect on system function (but not necessarily a total loss), and (3) a total loss of boundary integrity, which can be defined as a catastrophic loss of fluid and loss of function. Thus, in our example, a single function (preserve fluid boundary integrity) led to three distinct functional failures.

The key point in feature 2 is that we now make the transition to the equipment by identifying the *specific failure modes* in specific components that can potentially produce those unwanted functional failures.

Feature 3. In the RCM process, where our primary objective is to preserve system function, we have the opportunity to decide, in a very systematic way, just what order or priority we wish to assign in allocating budgets and resources. In other words, "all functions are not created equal," and therefore all functional failures and their related components and failure modes are not created equal. Thus, we want to *prior-*

itize the importance of the failure modes. This is done by passing each failure mode through a simple, three-tier decision tree which will place each failure mode in one of four categories that can then be used to develop a priority assignment rationale. (This will be discussed in detail in Chap. 5, Sec. 5.7.)

Feature 4. Notice that up to this point, we have not yet dealt directly with the issue of a preventive maintenance action. What we have been doing is formulating a very systematic roadmap that tells us the where, why, and priority with which we should now proceed in order to establish specific PM tasks. We thus address each failure mode, in its prioritized order, to identify candidate PM actions that could be considered. And here, RCM again has one last unique feature that must be satisfied. Each potential PM task must be judged as being "applicable and effective." *Applicable* means that if the task is performed, it will in fact accomplish one of the three reasons for doing PM (i.e., prevent or mitigate failure, detect onset of a failure, or discover a hidden failure). *Effective* means that we are willing to spend the resources to do it. Generally, if more than one candidate task is judged to be applicable, we would opt to select the least expensive (i.e., most effective) task. Recall that in Chap. 2, Sec. 2.3, when describing a run-to-failure task category, we indicated three reasons for such a selection. We can now be more precise, and state that failure of a task to pass either the applicability or effectiveness test results in two of the run-to-failure decisions. The third would be associated with a low-priority ranking and a decision not to spend PM resources on such insignificant failure modes.

In summary, then, the RCM methodology is completely described in four unique features:

1. Preserve functions.
2. Identify failure modes that can defeat the functions.
3. Prioritize function need (via the failure modes).
4. Select only applicable and effective PM tasks.

These four features or principles are actually implemented in a systematic, stepwise process that is described in detail in Chap. 5.

4.5   Some Cost-Benefit Considerations

As noted earlier, the primary driving force behind the invention of RCM was the need to develop a PM strategy that could adequately address system availability and safety without creating a totally impractical cost requirement. This has clearly been successfully achieved by commercial aircraft; however, quantitative data in the public arena on this cost picture is rather scarce. Figure 4.2 (Ref. 21) presents maintenance cost per flight hour in the first 10 years of RCM use. What we see in Fig. 4.2 is a maintenance cost that is essentially constant from the late 1960s to the early 1980s. This is precisely the period during which the 747, DC-10, and L-1011 were introduced into revenue service on a large scale. These jumbo jets not only introduced the large increase in passenger capability per aircraft (about 3 times over the 707 and DC-8), but also a higher daily usage rate and the deployment of several advanced technologies into everyday use. In spite of all of these factors, any one of which would normally tend to drive maintenance costs up, we see a fairly constant maintenance cost per flight hour historically occurring. RCM was the overriding reason for this.

Figure 4.3 (Refs. 9 and 21) presents another way to view the impact of RCM in the commercial aircraft world. Note that the PM definitions used in Fig. 4.3 correspond as follows to the PM task definitions given in Chap. 2, Sec. 2.3:

Hard-time   Time-directed
On-condition   Condition-directed
Condition-monitored   Run-to-failure

Two significant points can be observed with this data. First the pre- (1964) and post- (1969/1987) RCM periods reveal the dramatic shift that occurred in the reduction of costly component overhauls (i.e., hard-time tasks), mainly in favor of run-to-failure (i.e., condition-monitored) tasks. Much of this shift, of course, was made possible by a design

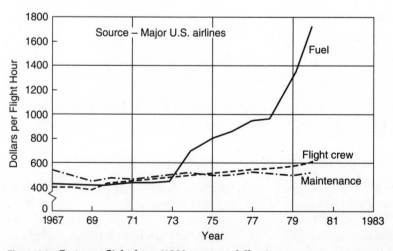

Figure 4.2 Costs per flight-hour (1982 constant dollars).

| Maintenance | Component distribution | | |
process	1964	1969	1987 (est.)
Hard-time* units	58%	31%	9%
On-condition† units	40%	37%	40%
Condition-monitored‡ units	2%	32%	51%

* *Hard-time*—Process under which an item must be removed from service at or before a previously specified time.

† *On-condition*—Process having repetitive inspections or tests to determine the condition of units with regard to continued serviceability (corrective action is taken when required by item condition).

‡ *Condition-monitored*—Process under which data on the whole population of specified items in service is analyzed to indicate whether some allocation of technical resources is required. Not a preventive maintenance process, CM *allows failures to occur*, and relies upon analysis of operating experience information to indicate the need for corrective action.

NOTE: Definitions from *World Airlines Technical Operations Glossary*—March 1981.

Figure 4.3   Commercial aircraft—component maintenance policy.

philosophy that included double and triple redundancy in the flight-critical functions. The RCM process was employed to take advantage of these design features in structuring where PM was critical and where the run-to-failure decision was appropriate. Also, notice that throughout the time period represented here, the condition-directed (i.e., on-condition) task structure remained fairly constant. The commercial aircraft industry was one of the early users of performance and diagnostic monitoring as a PM tool, and they have continued to successfully apply it throughout the generation of the newer jet aircraft.

The results indicated in Figs. 4.2 and 4.3 have led to a growing interest in other commercial areas. Most notably, nuclear power generation plants in several U.S. utilities are currently implementing RCM—or, at the very least, are conducting RCM pilot projects as a prelude to the conduct of a comprehensive RCM effort. Foreign nuclear utilities are also pursuing RCM, and Electricité de France (EDF) has in fact embarked upon a major effort to incorporate RCM as the basis for their PM program in all 50 operating nuclear units. Fossil power generation units in the United States have also started to apply RCM. In Chap. 7, selected information from RCM programs at GPU Nuclear Corporation, Florida Power & Light Company, and Westinghouse Electric Corporation will be presented to illustrate some typical RCM results. The reasons for this interest among the utilities and power plant operators are cost driven from two points of view:

1. Control and reduction of O&M costs

2. Increase in plant availability

The first RCM pilot study in 1984 at FP&L's Turkey Point Nuclear Plant provided the initial evidence of how an RCM program could favorably impact the O&M cost picture (Ref. 13). Analysis showed that a 30 to 40 percent savings in the existing PM program cost for the component cooling water system could be realized by implementing the RCM-based results. More recent surveys conducted by the Electric Power Research Institute (Ref. 22) have continued to quantify and verify this initial conclusion. In the EPRI survey involving seven different utilities, it was found that an average cost payback period of 6.6 years had been demonstrated with early RCM programs. This data was then extrapolated to a mature program state to show that the cost payback period could readily be reduced to 2 years or less—and this on the basis of PM cost savings alone. Other O&M benefits yet to be evaluated and credited to the cost-benefit picture include the following:

- plant trip reductions
- documented basis for the PM programs for use with regulators, insurers, and warranty contracts
- more accurate spare parts identification and stocking
- more efficient PM planning and scheduling
- decrease in corrective maintenance costs

But perhaps the most dramatic cost savings will be realized in the plant availability area where cost-avoidance benefits will be very large. For example, a base-loaded generation unit in the 1000 MW$_e$ range costs about $500,000–$750,000 for replacement electricity for *each* day of a forced outage. Thus, an RCM-based PM program that could raise plant availability by only one or two percentage points has a payoff in the multimillion-dollar range.

All of this is to say that the cost-benefit payoff with RCM has been dramatic in its impact on commercial aviation, and potentially offers similar dramatic payoffs in other areas where complex plants and systems are routinely operated.

5

RCM Methodology— The Systems Analysis Process

This chapter will provide a comprehensive description of the systems analysis process that is used to implement the four basic features which define and characterize RCM (see Chap. 4, Sec. 4.4). This process will be discussed in terms of seven steps that have been developed from experience as a most convenient way to systematically delineate the required information:

Step 1: System selection and information collection

Step 2: System boundary definition

Step 3: System description and functional block diagram

Step 4: System functions and functional failures

Step 5: Failure mode and effects analysis (FMEA)

Step 6: Logic (decision) tree analysis (LTA)

Step 7: Task selection

Satisfactory completion of these seven steps will provide a baseline definition of the preferred PM tasks on each system with a well-documented record of exactly how those tasks were selected and why they are considered to be the best selections among competing alternatives. A complete, albeit simple, systems analysis process using these seven steps is illustrated in Chap. 6 using a swimming pool as the example. Industrial examples, selected from RCM programs conducted on nuclear and fossil power plants and a USAF test facility, are given in Chap. 7. Various aspects of implementing the step 7 results are discussed in Chap. 8.

5.1   Some Preliminary Remarks

When an analyst embarks upon the process described in this chapter, it is helpful to keep a few key points in mind relative to the application of the RCM systems analysis process.

1. Traditional methods for determining PM tasks start with the issue of preserving equipment operability, and such methods tend to focus the entire task selection process on *what* can be done to the equipment. As a rule, *why* it should be done is never clearly addressed (or documented, if such consideration was, in fact, ever investigated). RCM is a major departure from this traditional practice! Its basic premise is "preserve function"—not "preserve equipment." This approach forces the analyst to systematically understand (and document) the system functions that must be preserved *without* any specific regard initially as to the equipment that may be involved. It then requires the analyst to think carefully about how functions are lost—in functional failure terms, not equipment failure terms. The purpose of this approach is to develop a credible rationale for why one might eventually desire to perform an appropriate PM task rather than just arbitrarily deciding to do something because "it sounds right." (The "preserve function" approach is initially developed in steps 3 and 4.)

2. However, this is not to imply that traditional experience and sound engineering judgment about equipment malfunctions is unimportant to the RCM process. To the contrary, the use of operations and maintenance personnel experience, as well as historical data from plant-specific and generic data files is an invaluable input to assuring that all important failure modes are eventually captured and considered in the FMEA (step 5).

3. The direct involvement of plant operations and maintenance personnel in the RCM systems analysis process is extremely important from another point of view also—namely, as a "buy-in" to the process. This embodies and promotes a feeling of belonging, and satisfies a very real necessity for them to share in the formulation of the PM tasks that they will eventually be asked to implement. Experience with several RCM programs has shown that success is rarely achieved if the buy-in factor has been neglected.

4. Recall from Chap. 2, Sec. 2.3 that there are four categories from which to choose a candidate PM task: (1) time-directed (TD), (2) condition-directed (CD), (3) failure-finding (FF), and (4) run-to-failure (RTF). As a rule, there is virtually no difficulty with people accepting the definition and use of TD and CD tasks in a PM program. Use of the FF task as a formal inclusion in the PM program is new to most people, but is generally accepted as a valid PM task in a short period of time. But the notion of a deliberate decision to run-to-failure is totally foreign to the more traditional elements of preventive maintenance, and frequently becomes a very difficult concept to sell. Thus, some care and sensitivity to the use of RTF tasks is necessary, and may entail some special education efforts to ensure that the operations and maintenance personnel understand why RTF tasks are, in fact, the best selection. The specific reasons behind RTF are developed in Steps 5, 6, and 7.

5. It is not uncommon for people receiving their first exposure to RCM to comment that "there sure is a lot of paperwork involved here." And in the framework, say, of a plant maintenance director and his or her staff, there is some truth in the comment. Thus, it becomes important to emphasize certain crucial points in order to help people to understand why the paperwork is there, and how it benefits them in the long run. These points should include the following:

- RCM wants to ensure that you can answer, both today and in the future, the "why" behind every task that will use your limited resources (i.e., preserve the most important functions). It is especially important to know why you may have elected *not* to follow an OEM recommendation since regulators and insurers often tend to hold them in high regard.

- RCM wants to ensure that your task selections derive from a comprehensive knowledge of equipment failure modes because it is at that level of detail where failure prevention, detection, or discovery must occur. If your task selection process, whatever it may be, does not do this, then there is no assurance that the task really does anything particularly useful.

- RCM wants to ensure that the most effective (least costly) task is chosen for implementation. Historically, this has not been done and, consequently, most PM programs fall far short of realizing the best return for the resources spent.

In order to realize these benefits, it does take some effort and documentation. But once a system has been through the RCM process, it produces a baseline definition of the PM program for that system which needs only periodic update to account for new information and system changes (see Chap. 8, Sec. 8.4). Thus, the systems analysis process is, in fact, a one-shot process that thoroughly documents where you are and why—a point of increasing concern in the current economic and regulatory climate. Further, as the RCM process has evolved and matured, much of the mechanics of the analysis has been computerized, and this has introduced efficiencies as well as eliminated the need for hard copy reports where such was desired. More is said on this point in Chap. 8, Sec. 8.2.

6. It should be noted that the RCM methodology focuses only on *what* task should be done and *why* (i.e., task definition). All tasks must likewise establish *when* the task should be done (i.e., task frequency or periodicity), but these intervals are derived from separate analyses that must consider and utilize combinations of company and industry experience to establish initial task frequencies. More sophisticated statistical tools may be employed when data is available to pursue this avenue; also, controlled measurement techniques known as *age exploration* can be used. More will be said on this in Chap. 8.

7. If possible, the systems analysis process should involve a team of two or three analysts. This will encourage not only cross-talk about what information should be included in each of the seven steps, but also a healthy level of challenge, questioning, and probing in that regard. Further, in such team arrangements, it is beneficial to include one team member who is *not* totally conversant in the system under investigation and one who is experienced. This will also help to develop the cross-talk and challenge process to the betterment of the end product.

All of these points will be addressed more fully in later discussions, but experience has shown that the analyst has placed himself/herself in a more proper state of mind to proceed with an RCM systems analysis if some initial appreciation of the above points has been acquired and accepted.

5.2   Step 1—System Selection and Information Collection

When a decision has been made to perform an RCM program at your plant or facility, two immediate questions arise:

1. At what level of assembly (component, system, plant) should the analysis process be conducted?

2. Should the entire plant/facility receive the process—and, if not, how are selections made?

Level of assembly.   For our discussion here, we can think of the following definitions to describe levels of assembly:

- *Part* (or *piece part*):   the lowest level to which equipment can be disassembled without damage or destruction to the item involved. Items such as microprocessor chips, gaskets, ball bearings, gears, and resistors are examples of parts. Notice that size is not a criterion in this regard.

- *Component* (or *black box*):   a grouping or collection of piece parts into some identifiable package that will perform at least one significant function as a stand-alone item. Often, modules, circuit boards, and subassemblies are defined as intermediate buildup levels between part and component. Pumps, valves, power supplies, and electric motors are typical examples of components.

- *System:* a logical grouping of components that will perform a series of key functions that are required of a plant or facility. As a rule, plants are composed of several major systems such as feedwater, condensate, steam supply, air supply, water treatment, fuel, and fire protection.

- *Plant* (or *facility*): logical grouping of systems that function together to provide an output (e.g., electricity) or product (e.g., gasoline) by processing and manipulating various input raw materials and feedstock (e.g., water, crude oil, natural gas, iron ore).

When PM planning is approached from the function point of view, experience has rather clearly shown that the most efficient and meaningful function list for RCM analysis is derived at the *system level*. In most plants or facilities, the systems have usually been identified, since they are also used as logical building blocks in the design process, and plant schematics and piping and instrumentation diagrams (P&IDs) thus define these systems rather precisely. These system definitions typically serve well as a starting point for the RCM process.

A reasonable way to explain and justify the use of systems in the RCM process is to consider the alternatives—that is, why not components on a one-by-one basis? Or, at the other extreme, why not the entire plant in a single analysis process? First, at the component level, it becomes difficult, sometimes impossible:

> to define the significance of functions and functional failures. For example, a valve can fail to open or fail to close on demand, and defeat some flow function that it controls; but unless the analysis looks more broadly at the system functions that are affected, we may not truly know the component function significance. We will also find later that a single component often supports several functions, and this becomes clear to the analyst only when viewing the entire system, not just one component; and

> to perform meaningful priority rankings between failure modes that are competing for limited PM resources. In a component we may have only two or, at most six to eight, failure modes to compare, whereas a system typically has hundreds, and comparisons make more sense.

At the other end of the spectrum, quite simply, the entire plant in one bite will literally choke the analysis process, and create an analysis nightmare in attempting to follow too many functions at once. Even combining two systems in one analysis (in a trial case, condensate and feedwater in a power generation plant) proved to be extremely cumbersome and difficult to track, and was abandoned in favor of two separate systems analyses before step 4 was completed. Generally, it can be stated that multiple systems analysis packages tend to exceed the cumulative time required to perform separate systems analyses due to the confusion created with the multiple system approach.

To summarize, then, the recommended approach is to conduct the RCM analysis process at the systems level—hence the term *systems analysis process.*

System selection.   Having established that the system is the best practical level of assembly at which to conduct the RCM analysis process, we can now focus on which systems to address and in what order. Obviously, one decision could be to treat all plant/facility systems. However, we have consistently found that such a course of action may not be cost-effective in that some systems have neither a history of failures nor excessive maintenance costs that might warrant a special investigation to "make it better." Given that such may be the case in your plant or facility, what procedure might be employed to select those systems with the highest potential for benefit from the RCM systems analysis process? Several selection schemes have been employed, and include the following:

1. Systems with a high content of PM tasks and/or PM cost

2. Systems with a large number of corrective maintenance (CM) actions over the past two years

3. A combination of item 1 and item 2

4. Systems with a high cost of CM actions over the past two years (this may indicate results different from item 2)

5. Systems with a large contribution to full or partial outages (or shutdowns) over the past two years

6. Systems with a high content of concern with respect to safety and environmental issues

It would appear from experience that any of these schemes, except item 6, can be used with approximately the same answers. In the case of item 6, safety and environmental issues, while real, are not necessarily good indicators of where maintenance improvements can or should be made.

In a typical selection process at the Florida Power & Light fossil power generation plants using item 5, a Pareto diagram such as that shown in Fig. 5.1 was constructed using Effective Forced Outage Rate (EFOR). Further, a threshold level of concern for EFOR was stipulated (say, a value of <0.10 in Fig. 5.1), and every system above the threshold was scheduled for inclusion in the RCM program, and every system below the threshold was dropped—or at least deferred until the higher-priority systems were completed. With the higher-priority systems, the order of analysis would be from left to right in Fig. 5.1 since the greatest benefit for outage reduction would be realized from those systems with the highest EFOR value. In the early RCM pilot studies at nuclear power plants, item 3 was employed (Refs. 13 and 14). Later, the RCM program at the Three Mile Island Unit 1 nuclear generating station initially used a combination of items 3 and 6, but found that safety issues were creating a bias toward systems that, maintenance-wise, were low-cost/low-problem systems. It then became necessary to recalibrate the selection process to include several additional factors such as:

- Deferring RCM on systems scheduled for significant modification until after such was completed

- Performing RCM prior to performance of safety system functional inspections

- Performing RCM on systems where management judgment indicated PM optimization benefits could be derived

Whatever method might be chosen, it is important that it be done as simply as possible, and without a large expenditure of time and resources. Generally, those systems deserving of an RCM program can quickly be pinpointed without any significant margin of error. As a point of calibration, nuclear power plants may choose 30 to 40 systems (of the 100 plus available) for the RCM process, and fossil power plants might choose 10 to 12 systems (of the 25 to 30 available) for the RCM process.

Information collection.   Considerable time and effort can be saved (and perhaps continuing frustration avoided) by researching and collecting, at the outset, some necessary system documents and information that will be needed in subsequent steps. A list of the documents and information typically required for each system by the RCM analysts are as follows:

1. System piping and instrumentation diagram (P&ID). See example in Fig. 5.2 (from Ref. 13).

2. System schematic and/or block diagram. Frequently, these are developed from the P&ID to help in the visual display of how the

system works, and usually are less cluttered than the P&ID, thus facilitating a good understanding of the main equipment and function features of the system. See example in Fig. 5.3.

3. Individual vendor manuals for the equipments in the system which will contain potentially valuable information on the design and operation of the equipments for use in step 5 (FMEA).

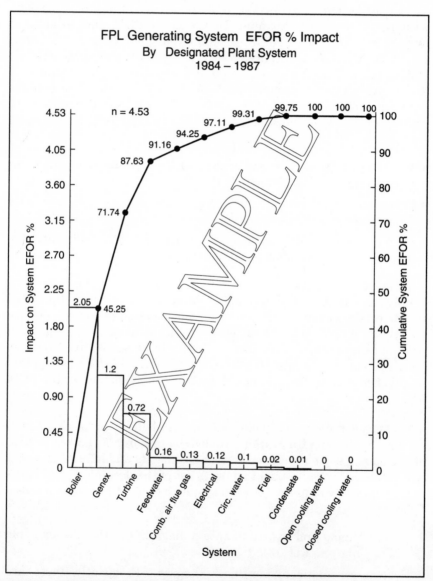

Figure 5.1 Typical EFOR Pareto diagram. (*Courtesy of Florida Power & Light.*)

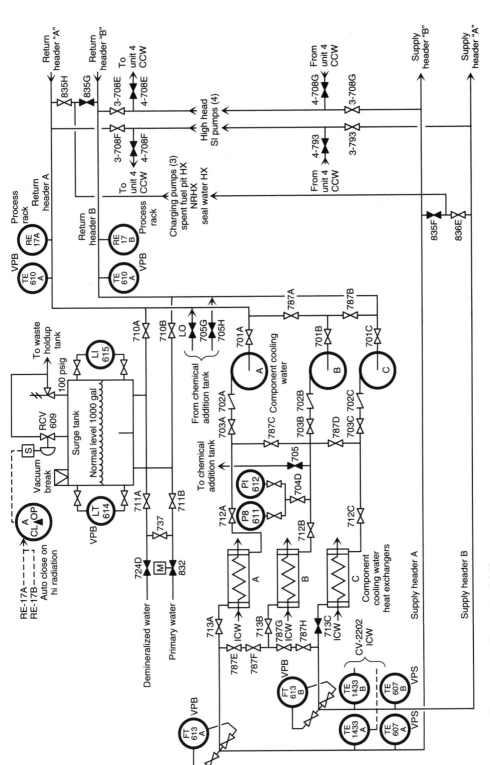

Figure 5.2 Component cooling water P&ID.

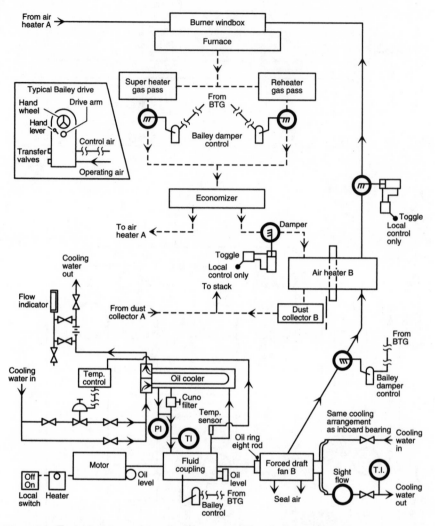

Figure 5.3 Typical system schematic. (*Courtesy of Florida Power & Light.*)

4. Equipment history files which will list the actual failures and corrective maintenance actions that have occurred in your facility for documentation in step 3 and for use in step 5.

5. System operation manuals which will provide valuable details on how the system is intended to function, how it relates to other systems and what operational limits and ground rules are employed. These items are of direct use in step 4 ("System Functions and Functional Failures").

6. System design specification and description data which will generally support and augment all of the preceding and, most importantly, will help to identify information needed in step 3 ("System Functional Description") and step 4.

There may be other sources of information that are unique to your plant or organization structure that would be helpful to accumulate. Also, industrywide data (such as equipment failure histories) are useful, when available, to augment your own experience. As you can see, all of the preceding items are aimed at assuring that the analyst has sufficient detail to thoroughly understand what is in the system, how it works, and what has been the historical equipment experience.

Suffice to say that you may not always have all of these items. For example, there may be no P&ID available, and you may have to "create" one via a system walkdown and visual reconstruction of the as-built configuration. Or you may have to conduct interviews to ferret out the equipment history. In older plants and facilities, this is frequently necessary. Even when the documentation is complete, system walkdowns and staff interviews are good ideas, and more will be said on these two points in later discussions.

One caveat: An item missing from the preceding list is the collection of documentation that defines the existing PM program on the system. This is eventually needed in step 7 ("Task Selection"), but it is *not* recommended that the analyst acquire the current PM program information until step 7 in order to preclude any prior knowledge on his or her part that might influence or bias the decisions on what the RCM results should be. There will be plenty of opportunity in the later stages of step 7 to collect this data and specifically compare it to the RCM results.

5.3 Step 2—System Boundary Definition

The number of separately identifiable systems in a plant or facility can vary widely depending upon plant or facility complexity, financial accounting practices, regulatory constraints, and other factors that may be unique to a given industry or organization. For example, in the power generation industry, an 800 MW$_e$ fossil (oil, gas, or coal-fired) plant typically will have about 25 to 30 separate systems, whereas an 800 MW$_e$ nuclear plant may well have in excess of 100 separate systems. Some gross system definitions and boundaries usually have been established in the normal course of the plant or facility design, and these system definitions have already been used in step 1 as the basis for system selection. These same definitions serve quite well to initially define the precise boundaries that must be identified for the RCM analysis process.

Why is precise system boundary definition so important in the RCM analysis process? There are two reasons:

1. There must be a precise knowledge of what has or has not been included in the system so that potentially important functions (and their related supporting equipments) are not inadvertently neglected—or, conversely, do not overlap with an adjacent system. This is especially true when RCM analyses are to be performed on two adjacent systems which, in all likelihood, will be done at different times and may involve different analysts.

2. More importantly, the boundaries will be the determining factor in establishing what comes into the system by way of power, signals, flow, heat, etc. (what we call the IN interfaces) and what leaves the system (the OUT interfaces). As will be discussed in steps 3 and 4, a clear definition of IN and OUT interfaces is a necessary condition to assuring accuracy in the systems analysis process. This, in turn, depends upon a clear understanding of what is or is not in the system. That is, where have the system boundaries physically been established?

There are no hard and fast rules that govern the establishment of system boundaries. Systems, by definition, usually have one or two top-level functions and a series of supporting functions that constitute a logical grouping of equipments. But considerable flexibility is allowed in defining precise boundary points to allow the analyst to group equipments in the most efficient manner for analysis purposes. Some examples of how this could occur will serve to best illustrate this latter point:

1. A heat exchanger may physically be in system A, but its level sensors are the key input to the control of flow in system B. Hence, the level sensors are placed in system B so that a complete picture of flow control in system B is possible.

2. An equipment lubrication function may reside in system A, but it ultimately services lubrication needs in several other systems. Here, it may be prudent to treat this lubrication function, in its entirety, as a completely separate System B.

3. System A may have control readouts in the plant control room that is physically far separated from system A. But the analyst may deem it best to include those control room instruments in his treatment of system A. Thus, if the control room is later analyzed as a separate system, the previously established boundary for system A would tell the analyst *not* to include those instruments in the control room boundary definition.

4. Other equipment items, such as circuit breakers (CB), can also be used as boundary points with the entire CB or only one side of the CB within the system boundary.

Whatever decisions are reached on boundary definitions, they must be clearly stated and documented as a part of the analysis process. Sometimes this is done by using forms such as are shown on Figs. 5.4 and 5.5, or by drawing the boundary lines on the system P&ID, process

RCM—Systems Analysis Process		
Step 2-1: System boundary definition		
Information: Boundary overview	Rev. no.: 0	Date: September 10, 1991
Plant: IEC Sayreville		Plant ID: IECS
System name: Condenser and air removal		System ID: ACC
Analysts: Smith, Worthy		

1. Major equipment included:

    Fans, tube modules, and all associated ductwork

    Vacuum deaerator and associated equipment

    Condenser receiver tank

    Hogging and two-stage holding air ejectors

2. Primary physical boundaries:

Start with:

    Steam turbine exhaust flange

    Outlet of steam bypass desuperheaters

    Inlet side of condensate makeup valves

    H.P. steam supply to air removal system

Terminate with:

    Condensate outlet of condenser receiver tank

    Air vent to atmosphere

3. Caveats of note:

    Air removal here refers only to condensate water delivered to suction side of condensate pumps (not deaeration of feedwater).

    This system does not include condensate pump.

    Condensate pump minimum flow considers only ejector condenser function, not condensate pump protection or gland steam condenser function.

Figure 5.4 System boundary definition—overview with typical example. (*Courtesy of Westinghouse.*)

RCM—Systems Analysis Process			
Step 2-2: System boundary definition			
Information: Boundary details	Rev. no.: 0	Date: September 10, 1991	
Plant: IEC Sayreville		Plant ID: IECS	
System name: Condenser and air removal		System ID: ACC	
Analysts: Smith, Worthy			

Type	Bounding system	Interface location	Dwg. or ref.
IN	Steam turbine	S.T. exhaust flange	PR-09-1001
IN	Main steam	#1 HRSG bybass connection at condenser ductwork	PR-10-1001
IN	Main steam	#2 HRSG bypass connection at condenser ductwork	PR-10-1001
IN	DCS	Input to level converter I/P LM01273 (2)	PR-10-1001
OUT	DCS	Output side of level transmitters LT01297 (3)	PR-10-1003
IN	Condensate	Inlet to condensate makeup manual-block valves 10-002 (2)	PR-10-1001
OUT	DCS	Output of amb. temp. elements TE-01003 & 01004 (2)	PR-10-1002
OUT	DCS	Output of condenser outlet TCs (16)	PR-10-1002
OUT	DCS	Output of vapor outlet TCs (4)	PR-10-1002
OUT	DCS	Output of fan motor vib. switches (16)	PR-10-1002
OUT	DCS	Output of condenser fan lube oil switch (16)	PR-10-1002
IN	DCS	Input to fan motor MCC011, etc. (16)	PR-10-1002
IN	480V AC power	Input to fan motor MCC011, etc. (16)	PR-10-1002
IN	DCS	Input to solenoid valve SV-01001/01002	PR-21-1002
IN	Instrument air	Output side of IAS header valve	PR-21-1002
OUT	DCS	Output side of vapor damper position switches (4)	PR-10-1002
IN	DCS	Input side of vacuum breaker solenoid	PR-10-1002
OUT	DCS	Output side of vacuum breaker position switch (2)	PR-10-1002
IN	Condensate	Inlet side of isolation & bypass valves for condensate min. flow valve 10-021 and 10-022	PR-10-1003

Figure 5.5 System boundary definition—details with typical example. (*Courtesy of Westinghouse.*)

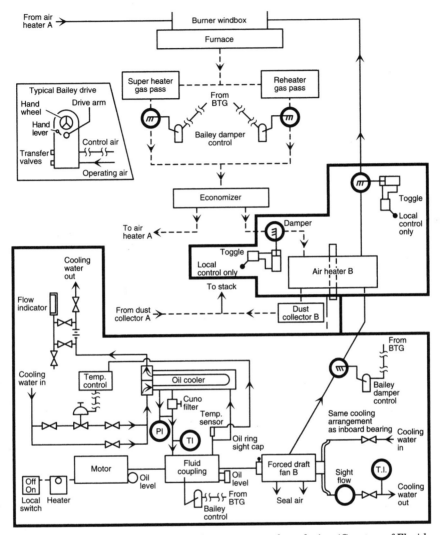

Figure 5.6 Typical system schematic indicating system boundaries. (*Courtesy of Florida Power & Light.*)

flow diagram, or the schematic block diagram (Fig. 5.6). Use of the latter approach in conjunction with the forms in Figs. 5.4 and 5.5 is the preferred method.

As the analyst proceeds with steps 3 and 4, it may become evident that the system boundary needs some adjustment to accommodate factors not originally envisioned. This is an acceptable practice, and in fact usually occurs as a part of an iteration process that will occur throughout steps 2, 3, and 4 in order to get the most efficient results before proceeding on to step 5.

5.4   Step 3—System Description and Functional Block Diagram

With our system selections complete, and the boundary definitions established for the first system to be analyzed, we now proceed in step 3 to identify and document the essential details of the system that are needed to perform the remaining steps in a thorough and technically correct fashion. Five separate items of information are developed in step 3:

- System description
- Functional block diagram
- IN/OUT interfaces
- System work breakdown structure
- Equipment history

System description.   By this point in the analysis process, a great deal of information has been collected and, to some degree, digested regarding what constitutes the system and how it operates. The analyst will now commit this information to the forms used in step 3 to document the baseline definition and understanding that is used to ultimately specify PM tasks. The first item of information is the system description that is documented on the form shown in Fig. 5.7. A well-documented system description will produce several tangible benefits:

1. It will help to record an accurate baseline definition of the system as it existed at the time of the analysis. Since design and operational changes in the form of modifications or upgrades can occur over time, the system must be baselined to identify where PM revisions might be required in the future (see Chap. 8, Sec. 8.4—"Living RCM Programs").

2. It will assure that the analysts have, in fact, acquired a comprehensive understanding of the system. (It is rare that the analysts are "experts" in more than two or three systems.)

3. Most importantly, it will aid in the identification of critical design and operational parameters that frequently play a key role in delineating the degradation or loss of required system functions. For example, cooling-water flow through a heat exchange may have several "allowable" inlet temperature and/or flow-rate conditions that correspond to varying degrees of "allowable" equipment cooling levels in successive states of system (or plant) capacity reduction short of a complete shutdown. Knowledge such as this becomes vital in the accurate specification of functional failures in step 4.

RCM—Systems Analysis Process		
Step 3: System description/functional block diagram		
Information: System description	Rev. no.:	Date:
Plant:		Plant ID:
System name:		System ID:
Analysts:		

1. Functional description/key parameters

2. Redundancy features

3. Protection features

4. Key instrumentation features (e.g., control)

Figure 5.7 System description form.

The level of detail found in system descriptions varies greatly from analyst to analyst. Suffice it to say that a well-documented form (Fig. 5.7) will pay significant dividends throughout the analysis process. Notice that Fig. 5.7 deliberately highlights the callouts for system redundancy features (e.g., on-line standby pumps, serial "normally closed" valves), protection features (e.g., alarms, interlocks, and trips) and key instrumentation features (e.g., control devices).

As an option, the analyst could elect to delay this system description task until the functional block diagram has defined the functional subsystems to be used. A decision could then be made to complete the system description in a single form at the system level, or to complete the description at the functional subsystems level on multiple forms. The latter is frequently done when the system in question is rather complex, and addressing it in discrete sections (i.e., the functional subsystems) is more manageable. Also, since the system analysis process from steps 4 to 7 is usually conducted on a functional subsystem level, it is often preferred to have the system descriptions likewise completed at the functional subsystem level.

Additional elements of the system description are further detailed and refined in the remaining four items in step 3.

Functional block diagram.   The functional block diagram is a top-level representation of the major functions that the system performs and, as such, the blocks are labeled as the functional subsystems for the system. As the name denotes, this block diagram is composed solely of functions; no equipments appear in this block diagram. Typical functional subsystems might include blocks for pumping or flow, heating, cooling, control, protection, storage, and distribution. Arrows connect the blocks to broadly represent how they interact with each other, and when finalized with the IN/OUT interfaces described next, give a complete *functional* picture of what our system is supposed to do. As you can probably envision at this point, the completed functional block diagram becomes a key link to step 4, where we will formally identify and document the system functions. The functional block diagram is recorded on the form shown in Fig. 5.8, which illustrates a typical diagram for a power plant turbine system—complete with all IN/OUT interfaces. This diagram, in addition to its value in helping us to visualize the system functional structure, will further subdivide the system into smaller packages for use in Steps 4 to 7. Frequently, this makes the analysis process less cumbersome, and even provides a logical basis for separating the work if more than one analyst is assigned to a given system. As a rule of thumb, we have found that systems should not be represented by more than five functional subsystems, and two or three such subsystems have become commonplace. It is rare that systems have more than five major functions—thus the reason for limiting the number of functional subsystems that should be used. When more than five are proposed, a close look will likely reveal that you actually have defined overlapping major functions. For example flow blockage is really a part of flow regulation, not a separate functional subsystem in its own right.

IN/OUT interfaces.   The establishment of system boundaries and the development of functional subsystems (i.e., major system functions) now permit us to complete a crucial part of the functional block diagram by observing and documenting the fact that a variety of elements cross the system boundary. These elements might include power, heat, signals, fluids, gases, etc. Some of these elements come "in" across the boundary from adjacent systems to service and assist in operating the functional subsystems, and some move "out" across the boundary to make other functions in the plant happen. These are called the *IN interfaces* and *OUT interfaces,* respectively. We can now see that the *OUT interfaces* are why the system exists, and thus they *will become the focus of the principle to preserve function.* In the systems analysis process, we assume that the *IN interfaces are always present and available when required.* True, these IN interfaces are required of our system, but the real product of the system is embodied in the OUT interfaces. Notice that IN interfaces here are OUT interfaces in some other system, so we are not really neglecting them. We document all of

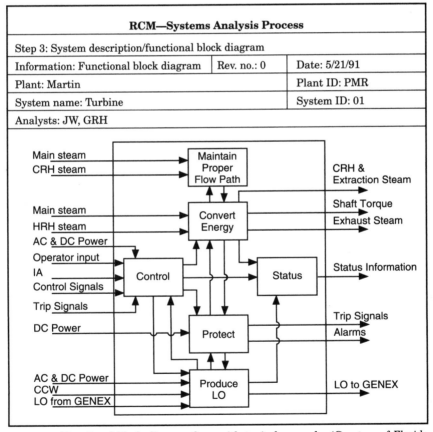

Figure 5.8 Functional block diagram form with typical example. (*Courtesy of Florida Power & Light.*)

this information on the form shown in Fig. 5.9, and schematically show it on the functional block diagram in Fig. 5.8.

When developing the functional block diagram, it is quite common to find that we must also include connecting lines between the functional subsystem blocks to represent the need for critical interactions within the system boundary. Note that Fig. 5.8 indicates several such interactions between the functional subsystems. In other words, functional subsystem A may have OUT interfaces that must take place for functional subsystem B to work properly—but this may all occur *within* the system boundary. We call these *internal OUT interfaces,* and are careful to identify them on Fig. 5.9, since they, too, can ultimately become a part of the functions that must be preserved. For example, steam may be an IN interface to functional subsystem A, and we assume it is

RCM—Systems Analysis Process		
Step 3: System description/functional block diagram		
Information: IN/OUT Interfaces	Rev. no.:	Date:
Plant:		Plant ID:
System name:		System ID:
Analysts:		
1. IN interfaces		
2. OUT interfaces		
3. Internal OUT/interfaces		

Figure 5.9 IN/OUT interfaces form.

there when needed. However, the steam continues on to functional subsystem B (that is, steam does not independently cross the boundary to serve subsystem B), and thus steam flow from A to B becomes an internal OUT interface for subsystem A. A lubrication function that resides totally within a system boundary is another example of a situation where internal OUT interfaces must be clearly identified to show the dependency between functional subsystems. Thus, internal OUT interfaces become important considerations in the analysis because system outputs that do cross the system boundary cannot successfully occur without them.

System work breakdown structure (SWBS). SWBS is a carryover from terminology that was used in Department of Defense applications of RCM, and is used to describe the compilation of the equipment (component) lists for each of the functional subsystems shown on the functional block diagram. Notice that this equipment list is defined at the component level of assembly (per the component definition described previously in Sec. 5.2). It is essential that *all* components within the system boundary be included on these equipment lists; failure to do so would automatically eliminate those "forgotten" components from any further PM consideration in steps 4 to 7. A correct P&ID can be used as an excellent source of information to develop an equipment list. In older plants or facilities, however, it is recommended that a system walkdown also be performed to assure the accuracy of the list.

Since most systems contain a sizable complement of instrumentation and control (I&C) devices, it is convenient to list I&C and non-I&C components separately under these respective headings. Later, we will further discuss some guidelines on how to efficiently handle I&C devices in the RCM systems analysis process, and a separate listing of the I&C devices will then be useful. The SWBS information is documented on the form shown in Fig. 5.10.

Equipment history. With the possible exception of new, state-of-the-art equipments, virtually all of the components on the SWBS have some history of prior usage and operational experience. For RCM purposes, the history of most direct interest is that associated with failures that have been experienced over the past 2 or 3 years. This failure history is usually derived from work orders that were written to perform *corrective* maintenance tasks. The equipment history information is recorded on the form shown in Fig. 5.11. Note that the primary information that we wish to capture is the failure mode and failure cause associated with the corrective maintenance action(s), since this information will be of direct value in completing step 5, the failure mode and effects analysis (FMEA).

RCM—Systems Analysis Process		
Step 3: System description/functional block diagram		
Information: System work breakdown structure	Rev. no.: 0	Date: 9/30/91
Plant: IEC Sayreville		Plant ID: IECS
System name: Condenser and air removal		System ID: ACC
Analysts: Smith, Worthy		

Cooling Subsystem

Noninstrumentation

Vapor ductwork (including expansion joints and rupture disk)
K type tube modules (12) + D (4)

Fan motors—K (12) + D (4)

Fan motor starter—K (12) + D (4)

Fan blade assembly—K (12) + D (4)

Fan gearbox—K (12) + D (4)

Flow valve FV01001 (for 86″ damper) (2)
        —including inflatable seal
Solenoid valve SV01001 (for 86″ damper) (2)
Air offtake header piping
        —condenser to condensate receiver tank
Drain piping—vapor ductwork to condensate receiver tank
Vacuum breaker FV01003
Solenoid valve SV01003 for vacuum breaker

Instrumentation

PT-01001	Pressure transmitter + isolation valve (3)
PT-01001	Pressure indicator + isolation valve (local) (1)
TE-01001	Temperature element (2)
TI-01001	Temperature indicator (local) (1)
PS-01010	Pressure switch on gearbox (16)
YS-01010	Vibration switch on gearbox (16)
ZS-01001	Position (limit) switch on 86″ damper (4)
	—2 open, 2 closed
TE-01011	Temperature element, condenser outlet @ 4/bank (16)
TE-01015	Temperature element, vapor outlet @ 1/bank (4)
TE-01003	Temperature element, ambient (2)

Figure 5.10 SWBS form with typical example. (*Courtesy of Westinghouse.*)

Where do we find this equipment failure history? First and foremost, if the system has already been in operation at the plant or facility in question, we should draw upon the plant-specific data that are available from the work order records or, if automated, the maintenance management information system (MMIS) files. In some instances,

RCM—Systems Analysis Process			
Step 3: System description/functional block diagram			
Information: Equipment history	Rev. no.:	Date:	
Plant:		Plant ID:	
System name:		System ID:	
Analysts:			
Component	Date	Failure mode	Failure cause

Figure 5.11 Equipment history form.

there may be sister plants or component usage in other similar facilities that are accessible from the same work order and MMIS files. Clearly, the in-house or plant-specific data are the most valuable since the records are reflective of operating and maintenance procedures that describe most accurately the actual components under investigation in the RCM analysis. In addition, there may be generic failure files that have been compiled on an industrywide basis that contain data of considerable value on the components in question. Frequently, these generic files may not contain the identical model or drawing number of interest to you, but with some care, components of a similar design may be applicable and useful to your analysis.

One cautionary note. The analyst should not be surprised to find that the equipment history files frequently contain very sketchy data on the failure event. Comments like "we found it broke—and fixed it" are not all that rare. This is unfortunate, not only because of its lack of useful data for your analysis, but also because it makes one wonder if the failure was ever really understood and thus fixed! Also, the data placed in the MMIS files may not be sufficient to meet the RCM analysis needs, and research back to the original work order may be necessary (if even possible at all). Even then, it is not uncommon to find a paucity of failure cause data, even when good failure mode data is present. In spite of this cautionary note, it is appropriate to search for component failure history in support of the FMEA in Step 5.

5.5 Step 4—System Functions and Functional Failures

The previous steps have all been directed toward developing an orderly set of information that will provide the basis for now defining system functions. This, of course, has all been done to satisfy the first RCM principal "to preserve system functions." It is therefore incumbent upon the analyst to define a complete list of system functions since subsequent steps will deal with this list in ultimately defining PM tasks that will "preserve" them. If a function is inadvertently missed, it is not likely that PM tasks directed at its preservation will be consciously considered!

It has already been noted in step 3 that the development of the OUT interfaces constitutes the primary source of information for system functions. In essence, every OUT interface should be captured into a function statement, and these function statements should be developed for each functional subsystem that has been previously defined. Other function statements can be derived from two additional sources:

1. There can be the internal OUT interfaces between functional subsystems that are essential to the successful performance of the system—yet they never cross the system boundary.

2. There can be passive functions which are required for successful system performance, but these usually do not explicitly appear as OUT interfaces. The most obvious passive functions are structural in nature—such as preserving fluid boundary integrity.

The analyst must keep in mind that these are *function* statements, not statements about what equipment is in the system. That is, avoid the use of equipment names to describe system functions. In some instances, however, reference to equipment or systems that are outside of the boundary are necessary to construct sensible function statements. Examples to illustrate correct and incorrect function statements are as follows:

Incorrect	Correct
Provide 1500 psi safety relief valves.	Provide for pressure relief above 1500 psi.
Provide a 1500 GPM centrifugal pump on the discharge side of header 26.	Maintain a flow of 1500 GPM at the outlet of header 26.
Provide alarm to control room if block valves are <90 percent open.	Provide alarm to control room if flow rate is <90 percent of rated value.
Provide water-cooled heat exchanger for pump lube oil.	Maintain lube oil ≤130°F.

When the system functions have been defined (again, for each identified functional subsystem), the analyst is ready to define the functional failures. That is, function preservation means avoidance of functional failures. We are now embarking upon the first step in the process of determining how functions might be defeated so that we can eventually ascertain the actions to prevent, mitigate, or detect onset of function loss. We need to keep two things in mind:

1. At this stage of the analysis process, the focus is on loss of function— *not* loss of equipment. Thus, as with the function statements, the functional failure statements are not talking about equipment failures (this will come in step 5).

2. Functional failures are usually more than just a single, simple statement of function loss. Most functions will have two or more loss conditions. For example, one loss condition may shut down an entire plant (a full, forced outage) while a less severe loss condition may result in only a partial forced outage or perhaps only some minor plant degradation. These distinctions are very essential so that ultimately the proper importance ranking can be determined in later portions of the analysis process (not all functional failures are equally important). In addition, these distinctions often lead to different modes of failure in the equipment that supports them, and this needs to be identified in step 5.

Let's illustrate this discussion with a couple of the preceding function statements.

Function	Functional failure
1. Provide for pressure relief above 1500 psi.	a. Pressure relief occurs above 1650 psi. b. Pressure relief occurs prematurely (below 1500 psi).
2. Maintain a flow of 1500 GPM at the outlet of header 26.	a. Flow exceeds 1500 GPM. b. Flow is less than 1500 but greater than 1000 GPM. c. Flow is less than 1000 GPM.

In function 1, a functional failure occurs if the pressure relief is greater than the 10 percent margin in the design (not precisely at 1500 psi) where pipe rupture would occur, but anything below 1500 psi essentially drains the system of its fluid. In both cases, the system is totally lost. In function 2, excessive flow might violate a system design condition and destroy some chemical process, whereas flow as low as 1000 GPM can be tolerated with some output penalty. But if flow drops below 1000 GPM, the system must be shut down.

You have probably noticed that an accurate portrayal of functional failures relies heavily on the design parameters of the system. For example, in function 2, it was necessary to understand that there was some allowable tolerance in the flow rate before a total shutdown and complete loss of process control occurred. This range-of-conditions situation is actually a fairly common occurrence, so the analyst must be careful to ensure that the functional failures completely describe the intended design conditions for each system function. It is rare that a function is either go or no-go. Recall that this point was emphasized earlier (in Sec. 5.4) when dealing with the development of information for Fig. 5.7, the system description form. We record the function and functional failure information on the form shown in Fig. 5.12.

5.6 Step 5—Failure Mode and Effects Analysis

Functional failure–equipment matrix. Step 5 now brings us to the question of which system equipments could play a role in the creation of a functional failure—or, stated somewhat differently, which equipments have the potential to defeat our principal objective to "preserve function." This is the first time in the systems analysis process that we directly connect the system functions and the systems equipments. This is done by completing the matrix shown in Fig. 5.13. Notice that the matrix relates functional failures (not functions) to the equipments since it is the functional failure that we must strive to avoid in the preventive maintenance actions that we are seeking to define. This matrix is developed for each of the functional subsystems that were previously

RCM—Systems Analysis Process		
Step 4: Functions/functional failures		
Information: Functional failure description	Rev. no.:	Date:
Plant:		Plant ID:
System name:		System ID:
Analysts:		

Function no.	Functional failure no.	Function or functional failure description

Figure 5.12   Function/functional failure form.

delineated. The respective functional failures and their equipments
(components) as delineated in the system work-breakdown structure
then form the horizontal and vertical elements of the matrix for each
functional subsystem. The analyst's task at this point is to identify, for
each functional failure, those components which could play a role in
it—and to so indicate by placing an X in the appropriate intersection
box. Clearly, this task requires a reasonable knowledge of the system
design and operation characteristics, and the analyst should not be

bashful in seeking assistance from engineering and operations specialists in completing the matrix. We usually find that some errors or
omissions are made in completing the matrix on the first go-around,
and thus should not be surprised if subsequent efforts with the FMEA
require that some adjustments be made. When the matrix is completed, we will have developed the specific road map to guide us hence-
forth in the system analysis process.

A word of caution is in order here relative to the frequently used
notion of "critical components." In many areas of system and plant
analysis, analysts often conduct an exercise aimed at identifying the
critical components. The underlying thought here, of course, is that
there must be some *noncritical* components in the system or plant that
we don't really have to pay too much attention to, and once they are
identified we can, in fact, discard them from any further consideration.
Were this true, certain resources could be saved. A variety of methods
are employed to identify these noncritical components; most of them
are qualitative and highly subjective in nature and, as a rule, never
directly consider the function(s) that are supported by these components. It is the author's view that this type of exercise is virtually
worthless, and frequently leads to the exclusion of many important
equipments from appropriate attention. My view is that what we really
have are critical *functions,* and that these functions are not all created
equal (we will separate them in a priority ranking in step 6). However,
we should be very careful not to prematurely discard components as
noncritical until we have truly identified their proper tie and priority
status to the functions and functional failures. In the matrix just
described in Fig. 5.13, we have taken the first crucial step to do just
that. It may surprise you to find that only on rare occasions will we
have an empty set in Fig. 5.13. That is, an item of equipment will have
no X mark at any functional failure intersection. When, and if, this
occurs, we have either made a mistake in our analysis, or we have discovered a component that plays absolutely no useful role in the functional subsystem. (The author has seen the latter happen twice, and
action was taken to remove or block out the component.) Otherwise, all
components are "critical" in that they can play a role in creating one or
more functional failures. The only question that remains is how one
should prioritize that role in a world where components will ultimately
compete for PM resources. And this will be answered in step 6. But to
prematurely discard any component as "noncritical" without understanding its relationship to the functions and functional failures is a
very dangerous course to pursue. The one exception to this is discussed
following.

And now a brief word about instruments. First, systems usually
have a large number of instruments. Second, they can be easily cate-

RCM—Systems Analysis Process

Step 4: Functions/functional failures		
Information: Equipment–functional failure matrix	Rev no.:	Date:
Plant:	Plant ID:	
System name:	System ID:	
Analysts:		

No.	Equipment (or component) name	Functional failures								

Figure 5.13 Equipment–functional failure matrix.

gorized as: (1) perform control functions, (2) provide alarms, and (3) provide status information only. You may wish to take the instruments on the SWBS, and so indicate one of these three categories for each. It is the author's recommendation that those instruments categorized as "status information only" be *dropped* from any further consideration in the system analysis process, and put on the run-to-failure list. Quite simply, such instruments are always a very low PM consideration (or should be), and here is the proper place in the RCM process to so indicate. Remember, you can always change your mind later (in Step 7) when you perform a final sanity check on all run-to-failure items.

The FMEA. The Failure Mode and Effects Analysis will utilize the form shown in Fig. 5.14 and will follow the general form of analysis that was first described in Chap. 3, Sec. 3.5. The analyst will progressively march through the functional failures for each functional subsystem, and for each functional failure will sequentially address each component that was identified in Fig. 5.13 as having some potential role in the functional failure under investigation. As noted previously, at this level of detailed analysis, there could be some modification either to add or delete components in the matrix in Fig. 5.13.

While each step in the systems analysis process is important to achieving a comprehensive and accurate PM program, we cannot overemphasize the significance of the FMEA. It is at this point that the analyst will define the *specific component failure modes and causes* that can defeat the system functions—and it is only at this level of detailed knowledge that we can finally determine if some appropriate PM task can prevent, mitigate, or detect the onset of that failure mode. Most PM programs in place today fail to recognize this fundamental fact, and thus specify PM tasks without a complete understanding of why it should be done! That is to say that the current PM mind-set fails to carry its analysis and selection to the failure mode/failure cause level, and thus PM task selection tends to be more of a guessing game than a logically developed choice. If you, the reader, are an experienced maintenance engineer or technician, you will admit the validity of this observation, and will undoubtedly be thinking back over your experience and identifying PM tasks about which you now say, "Why did we ever do that one?"

Let's discuss the guts of the FMEA process by following the columns on Fig. 5.14. First, we shall have one or more sheets devoted exclusively to each functional failure, and this is identified in the header data along with the other standard descriptors. Then, for each component listed in the matrix on Fig. 5.13, the analyst must first establish specifically *how the component must fail* in order to produce the func-

RCM—Systems Analysis Process

Step 5: Failure mode and effects analysis	Rev no.:	Date:
Functional failure no.:	FF title:	
Plant:	Plant ID:	
System name:	System ID:	
Analysts:		

Component	Failure mode	Failure cause	Failure effect			
			Local	System	Plant	LTA

Figure 5.14 FMEA form.

tional failure in question. This is the *failure mode,* and we describe it in four words or less if at all possible. Figure 5.15 presents a partial list of terms commonly used to describe failure modes. Thus, if we state a valve is "jammed closed" or a cable is "insulation cracked off," we are describing *what* the PM task will have to address (if it can). Please note that many components will have more than one failure mode associated with any given functional failure and, clearly, it is necessary to identify all of them. Again, the analyst is confronted with the requirement to now understand specific components to the point that he or she can credibly define these failure modes. There are at least three sources of information to help in this process. First, the equipment history file that was developed in step 3 can tell us something about the failure modes that have actually occurred on the component. However, we do not limit our analysis to only actual failure mode occurrence but, rather, we extend it to include all *plausible* failure modes irrespective of their perceived probability of occurrence. However, in order to maintain a proper level of reality in the analysis, we also recognize that some failure modes are implausible, and we should strive to avoid letting them needlessly clutter the process. Examples of implausible failure modes could include the following:

- "Inadvertent mechanical closure" of a normally open *manual* valve
- "Structural collapse" of ductwork in a benign environment
- "Rupture" in low pressure pipe

abrasion	damaged	lack of —	ruptured
arcing	defective	leak	scored
backward	delaminated	loose	scratched
out of balance	deteriorated	lost	separated
bent	disconnected	melted	shattered
binding	dirty	missing	sheared
blown	disintegrated	nicked	shorted
broken	ductile	notched	split
buckled	embrittlement	open	sticking
burned	eroded	overheat	torn
chafed	exploded	overtemp	twisted
chipped	false indication	overload	unbonded
clogged	fatigue	overstress	unstable
collapsed	fluctuates	overpressure	warped
cut	frayed	overspeed	worn
contaminated	intermittent	pitted	
corroded	incorrect	plugged	
cracked	jammed	punctured	

Figure 5.15 Typical descriptors for failure modes.

The second information source is the experience that can be gleaned from engineers, technicians, and senior craftpeople who have direct design and hands-on experience with the components. More often than not, this level of assistance is found at the plant or facility under investigation. The analyst should go out of his or her way to involve these plant personnel; they will not only provide expert help, but their direct involvement in the FMEA process will begin to develop their *buy-in* to the RCM program. This buy-in will ultimately be a necessary ingredient in obtaining their wholehearted support to implement the RCM results from step 7.

A third information source may be the original design FMEA that was done by the OEM, or documented collections of failure mode information such as that contained in Ref. 23.

In the next column, we attempt to identify the *root cause* of each failure mode. The root cause refers to the basic reason for the failure mode—that is, *why* the failure mode occurred. The root cause can always be *directly* identified with the failure mode and component in question—as opposed to the consequential cause, which refers to some component failure elsewhere in the system which *indirectly* caused the failure mode. Consequential cause is of no interest in a maintenance analysis because no amount of maintenance on component B will ever help to avoid a failure in component A. For example, a pump motor may have seized bearings as a failure mode due to lack of proper lubrication, but this is a consequential cause resulting from failure in a separate lube oil system that feeds several items of equipment. No amount of PM on the motor bearings will prevent a basic failure mode of, say, a clogged filter in the separate lube oil system. However, bearing seizure due to contamination buildup in a self-contained oil reserve is a root cause that can only be addressed at the motor itself. The analyst will undoubtedly find that equipment history files are quite sparse when it comes to root cause information. This again points to a glaring deficiency in data systems. (If you don't know why a failure occurred, how can you be sure it's fixed?) But this is a fact of life, and the author's advice here is to do your best to intelligently select one or two likely root causes for entry onto the form. The reason for my emphasis on attempting to establish a root cause, even if only "guesstimated," is that this piece of input may eventually prove crucial in selecting a candidate PM task. In addition, some (less than 5 percent) failure modes have two credible root causes for a single failure mode, and could conceivably require two different PM tasks for the same failure mode!

The final step in the FMEA process is the effects analysis portion of the form. Here, the analyst will determine the *consequence* of the failure mode, and this will be done at three levels of consideration—locally, at the level of the component in question; at the system level;

and, finally, at the plant level. There are two primary reasons for conducting the effects analysis at this point: (1) we want to assure ourselves that the failure mode in question does in fact have a potential relationship to the functional failure being studied and (2) we want to introduce an initial screening of failure modes that, by themselves, cannot lead to a detrimental system or plant consequence. In order to fully understand the significance of these two statements, we need to introduce and discuss the *single failure rule* and how we treat redundancies that have been designed into the system.

Redundancy—general rule.   Our objective in RCM is to preserve function. Thus, in the maintenance strategy of how we view the commitment of resources, it becomes important to first commit those resources to *single* failure occurrences that detrimentally impact function. If redundancy prevents loss of function, then a failure mode thus shielded by redundancy should not be given the same priority or stature of a failure mode that can singly defeat a necessary function. Note that if one is truly concerned that there is a high probability of multiple independent failures in a redundant configuration, then what you have identified is a more fundamental design issue, and not one that should be addressed or solved by the maintenance program.

So, how do we invoke this redundancy rule? Quite simply, when listing the failure modes, we do *not* introduce the redundancy rule since our objective is to assure that we initially capture each failure mode (protected or not) that can lead to the functional failure. But then, in the effects analysis, we apply the redundancy rule. If available redundancy essentially eliminates any effect at the system level (and, it will follow, at the plant level also), we drop the failure mode from further consideration, and place it on the run-to-failure (RTF) list that will receive a further review and sanity check in step 7. Since complex plants and facilities are often designed with a host of redundancy features in order to achieve high levels of safety and productivity, it is not uncommon to find that this initial screening with the redundancy rule could relegate 50 percent or more of the failure modes to the RTF status. Should you encounter this situation, your maintenance program will likely realize significant cost reductions from the foresight that occurred during the design phase (even though that foresight was not, in all likelihood, maintenance-driven).

Redundancy—alarm and protection logic.   There is one important exception to the preceding rule, which involves alarms and protection logic devices involving some voting scheme. Here, the rule requires an assumption of multiple failures in order to properly assess the effects or consequences of alarm or protection loss. In the case of alarms, a "failure to operate" is, by itself, not significant. It can become significant, however, if the alarmed component is also failed. So, we assume that the alarmed component is failed in order to place the proper perspective in the effects analysis on the consequence of not knowing that such has occurred. The same principle holds with protection logic where the redundant channels are assumed failed to the extent that the next single failure will wipe out the protection logic. We tend to find protection logic systems when dealing with safety and environmental issues or areas where "trips" must occur automatically to preclude widespread damage to a plant.

Again, if the system effect (and thus the plant effect) is "none" as a result of applying the redundancy rule, we drop the failure mode from further consideration in the analysis until the final sanity check. Conversely, when there is some form of system and/or plant effect, we retain the failure mode for further consideration. When there is a choice to be made from several possible failure effects, we always choose the worst-case scenario in order to reflect the most severe consequence that could result from the failure mode. Such choices could occur, for example, as a function of time of occurrence (start-up, steady state, etc.) or plant operating parameters (flow rate, pressure temperature, etc.). The last column on the FMEA form, labeled *LTA* or logic tree analysis, is where we signify a *yes* or *no* to indicate whether or not the failure mode will be carried forth to step 6—the logic (decision) tree analysis (LTA). The step 5 process is continued until each functional failure and its related components have been through the FMEA.

5.7 Step 6—Logic (Decision) Tree Analysis (LTA)

The failure modes that survive the initial screening test in the effects analysis in step 5 will now be further classified in a qualitative process known as the *logic tree* or *decision tree analysis* (LTA). The purpose of this step is to further prioritize the emphasis and resources that should be devoted to each failure mode, recognizing as we have earlier that all functions, functional failures and, hence, failure modes are not created equal.

Several ranking schemes could conceivably be used to achieve a priority listing of the failure modes, but the RCM process uses a simple three-question logic or decision structure that permits the analyst to quickly and accurately place each failure mode into one of four categories (or *bins* as we often call them). Each question is answered either *yes* or *no*. As we shall see momentarily, the bins form a natural importance ordering to the failure modes.

The basic LTA uses the decision tree structure shown in Fig. 5.16. The information that is gathered from this tree is recorded on the form shown in Fig. 5.17. You will notice that this decision process will identify each failure mode in one of three distinct bins: (1) safety-related, (2) outage-related, or (3) economics-related. It also distinguishes between evident (to the operator) or hidden. Let's examine the details of how the LTA is used.

Each failure mode is entered into the top box of the tree on Fig. 5.16, where the first question is posed: Does the operator, in the *normal* course of his or her duties, know that something of an abnormal or detrimental nature has occurred in the plant? It is not necessary that the operator know exactly what is awry for the answer to be *yes*. The

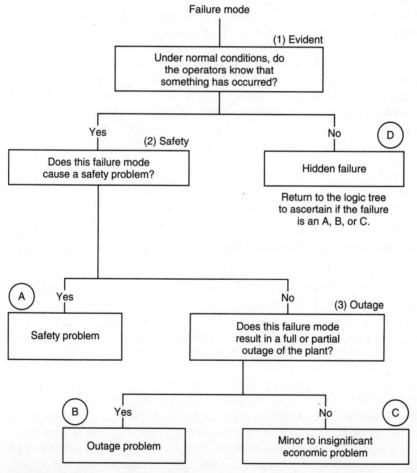

Figure 5.16 Logic tree analysis structure.

RCM—Systems Analysis Process

Step 6: Logic tree analysis

Information: Failure mode criticality	Rev no.:	Date:
Plant:	Plant ID:	
System name:	System ID:	
Analysts:		

Functional failure	Component and failure mode	Criticality analysis				Comment
		Evident?	Safety?	Outage?	Category	

Figure 5.17 Logic tree analysis form.

reason for this question is to establish initially those failure modes that may be hidden from the operator. Failures in standby systems or components are typical of hidden failures; unless some deliberate action is taken to find them, they will not be discovered until a demand is made, and then it may be too late. Thus, hidden failures could later give rise to failure finding PM tasks. Evident failures, however, alert the operators to act, including taking the necessary steps to detect and isolate the failure mode if such is not immediately visible. So a *yes* to the first question leads us to the next question in the tree while a *no* leads us directly to bin D—or the hidden function bin.

Evident failure modes now pass to the second question, which asks if they can lead to a safety problem. Safety, in the context used here, refers to personnel death or injury, either on-site or off-site. However, you can define safety in whatever fashion your particular needs may dictate. For example, safety may be limited to include only off-site injuries or deaths; or safety may be defined to include violation of EPA standards or even equipment damage. The author's preference is to limit safety to personnel injury or death, but this is strictly a personal choice. When broadened beyond this, there may be different levels of safety that must be classified. In any event, if the second question yields a *yes*, the failure mode is placed in bin A—or the safety bin. A *no* takes us to the third and final question.

If there is no safety issue involved, the remaining consequence of interest deals solely with plant or facility economics. Thus, the third question is formulated to make a simple split between a large (and usually intolerable) economic penalty, and a lesser (and usually tolerable for at least some finite time period) economic penalty. This is done by focusing on plant outage or loss of productivity. The question becomes: Does the failure mode result in a loss of output ≥ 5 percent? This can also be stated as: Does the failure mode result in a full or partial plant outage (where partial can be defined as ≥ 5 percent)? The selection of the 5 percent threshold value depends upon several variables, so the analyst should adjust this value to suit the situation at hand. A *yes* answer puts us in bin B, which is the outage bin, and signifies a significant loss of income as the consequence. As an example, a full outage in a base-load 800 MW_e electric power plant can cost upwards of $750,000 per day to purchase replacement power! A *no* answer tells us that the economic loss is small and places us in bin C. That is, the failure mode is essentially tolerable until the next target of opportunity arises to restore the equipment to full specification performance. There are many examples of bin-C-type failure modes, and these include items such as small leaks and degraded heat transfer due to tube scaling.

Let's return for a moment to bin D—hidden failure modes. We must additionally classify each hidden failure as safety-, outage-, or economic-related in order to know the potential severity involved. When the LTA process is concluded, every failure mode that was passed to the LTA will have been classified as either A, B, C, D/A, D/B, or D/C.

What do we do with this information? We use it, if you will, to separate the wheat from the chaff. In a world of finite resources, in other words, who gets the favorable nod? I am sure, by now, that failure modes, either evident or hidden, which land in bin A or bin B would have your priority over bin C. And, in general, bin A has the priority over bin B. The current litigation environment, if nothing else, makes that choice an easy one. So, we usually choose to address PM priorities as:

1. A or D/A
2. B or D/B
3. C or D/C

Bin C, in particular, tends to raise a dilemma in the sense that its potential consequence is small, by definition—but we hate to just walk away from those failure modes. While each case must be viewed on its own merits, we should note that the evidence is rather strong that bin C should be relegated to the run-to-failure list without further ado. If this should be done, you will notice that the accumulation of RTFs from the "status only" instruments, the effects analysis in step 5, and now the RTFs from bin C in step 6 could sum to a sizable list! In most instances, the sanity check on these failure modes will leave them in their RTF status. This list alone, at this point in the systems analysis process, could constitute a sizable reduction in O&M costs if all current PM tasks which apply to these failure modes were eliminated! It is the author's recommendation that all bin C failure modes be designated as RTF, and changed only if they should not pass the sanity check in step 7. In this case, only failure modes that have been assigned an A or B classification would be passed on to step 7.

5.8 Step 7—Task Selection

The task selection process. Our systems analysis efforts to this point have been directed to delineating those failure modes where a PM task will give us the biggest return for the investment to be made. So, for each of these failure modes, our job now is to determine the list of applicable candidate tasks, and then to select the most effective task from among the competing candidates. Recall from Chap. 4, Sec. 4.4

that PM task selection in the RCM process requires that each task meet the *applicable and effective* test which is defined as follows:

- *Applicable.* The task will prevent or mitigate failure, detect onset of failure, or discover a hidden failure.
- *Effective.* The task is the most cost-effective option among the competing candidates.

If no applicable task exists, then the only option is RTF. Likewise, if the cost of an applicable PM task exceeds the cumulative costs associated with failure, then the effective task option will also be RTF. The exception to this rule would be a bin A, or safety-related, failure mode where a design modification may be mandatory.

Developing the candidate list of PM tasks is a crucial step, and frequently requires help from several sources. Again, involvement in task selection from the plant maintenance personnel is necessary to realize the benefit of their experience as well as to gain their buy-in to the RCM process. However, other sources of input—such as operations personnel, technical data searches, and vendor expert advice—are recommended to assure the inclusion of state-of-the-art technology and techniques. This latter statement is especially true regarding the introduction of performance monitoring and predictive maintenance options for CD tasks.

The road map in Fig. 5.18 and the form in Fig. 5.19 are used to structure and record the task selection process. The road map, in particular, is very useful in helping the analyst to develop logically the candidate PM tasks for each failure mode. Briefly, the steps in Fig. 5.18 are as follows:

1. We have previously discussed the significance associated with a knowledge of an equipment's failure density function (Chap. 3, Sec. 3.4), and the danger that befalls us should we erroneously select overhaul (or TD) tasks if we do not know the failure density function (Chap. 4, Sec. 4.2). Thus, this first question requires that we acknowledge how much we really know about the equipment age-reliability relationship (i.e., the failure density function). In all likelihood, we rarely know it precisely but, more frequently, we may have some reasonable estimate of it and can answer *yes* or *partial.* But in the absence of at least a reasonable estimate, we must answer *no,* and not fall into the trap of trying to search for an applicable TD task which could actually make things worse—not better.

2. When we do have the age-reliability information (or least some partial evidence of such), this signals that we do understand (or can conservatively estimate) both the mechanisms/causes associated with

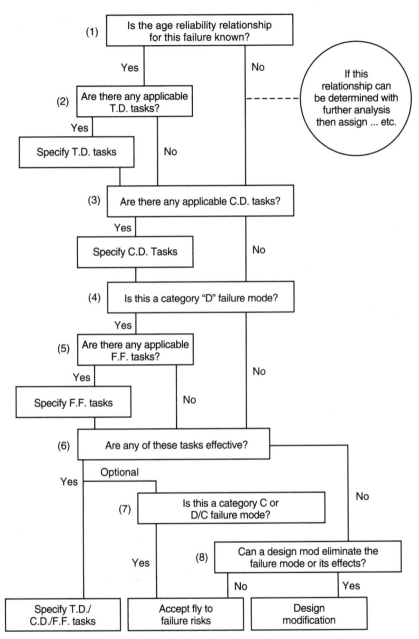

Figure 5.18 Task selection road map. (*Developed in cooperation with Mississippi Power.*)

RCM—Systems Analysis Process

Step 7: Task selection

Information: Selection process and decision	Rev no.:	Date:
Plant:	Plant ID:	
System name:	System ID:	
Analysts:		

FF no.	Component and failure mode	Failure cause	Selection guide								Candidate tasks	Effective-ness info.	Sel. dec.	Est. freq.
			1	2	3	4	5	6	7	8				

Figure 5.19 Task selection form.

the failure mode, and how the failure rate deteriorates over time. In other words, we know what task to select to prevent the failure mode and also when it should be done to minimize the chance of its occurrence. Note that if the age-reliability information shows a constant failure rate over the total expected life of the equipment, there is no applicable TD task available to us because any failure occurrence is strictly random in nature.

3. Even if a candidate TD task has been defined, we will further explore the possibility of an applicable CD task(s). Frequently, this is a smart path to pursue if our age-reliability information (and thus TD task selection) is on somewhat shaky ground. An appropriate CD task, aimed at measuring some telltale parameter over time, may well be the best selection ultimately. It will help us to develop the age-reliability information and, until then, will give us a high confidence in taking preventive measures (usually overhaul in nature) at the correct time. Of course, if your plant mimics the data in Chap. 4, Fig. 4.1, the CD task may well tell you that you never see the onset of the failure mode during the useful life of the equipment! Hopefully, if the answer to question 1 was *no,* you will find at least one candidate task in question 3. Don't be surprised, however, if you do not find either a TD or CD candidate task, since some failure modes are just not amenable to a PM action, even when age-reliability information is fully known.

4. Going back to the LTA information, is this a hidden failure mode?

5. If *yes* to question 4, can we specify a candidate FF task? In all likelihood, there will be some type of failure finding that can be considered. It is rare that some form of failure finding test or inspection cannot be done. When an FF task is selected, we would define its frequency such as to eliminate or significantly minimize any system or plant downtime that might be required to correct the failure.

6. Now we are ready to examine the relative costs associated with each candidate task, and this always includes the option of RTF costs. The job here is to select the lowest cost option. You may wish to refer to Chap. 2, Sec. 2.3 where the discussion on failure finding, using the auto spare tire example, illustrates how it is possible to have applicable TD, CD, and FF tasks with the ultimate selection being based on the effectiveness measure.

7. In step 7, we considered previously the relegation of all bin C and D/C failure modes to an RTF status. If this has not already been done, question 7 again asks that we consider such action.

8. This question is aimed at directing the analyst to consider design modification as a solution when no applicable and effective task has been identified. In the case of bin A or safety-related failure modes, consideration of a design modification should be *mandatory,* and presented to management for final decision on what should be done.

The form in Fig. 5.19 is used to record all of the decisions that were made during the task selection process, including the final selection which is recorded in "Sel. dec." column. The last column, "Est. freq.," is where we will record the suggested frequency or interval that should be assigned to the task. Some further remarks on the subject of task periodicity may be found in Chap. 8, Sec. 8.3.

Task comparison. If your RCM application is to an existing plant or facility, then there is a PM program of one sort or another already in place. One of several issues may be motivating management to upgrade this existing program. But, in so doing, it is fairly certain that management will want to know how the RCM-based PM tasks stack up against the current PM tasks. How different is the RCM program, and what is the nature of those differences? Even in a new plant or facility, it may be very important to compare the OEM recommendations to the RCM-based PM tasks. The form in Fig. 5.20 is used to collect such comparison information.

The selected PM tasks from the task selection form (Fig. 5.19) are listed in the "RCM-based task description" column in Fig. 5.20 and their components and failure modes of origin are likewise listed for traceability purposes. Since there is probably no established PM task-failure mode relationship in the existing program, the analyst can only try to match current tasks with the RCM tasks to see where they may be alike. Any task that does not match is then listed in the "Current task description" column with no counterpart RCM task listed at all. When the form is complete, it will contain four distinct categories of comparison:

1. RCM-based and current PM tasks are *identical.*
2. Current PM tasks exist, but should be modified to meet the RCM-based tasks.
3. RCM-based PM tasks are recommended where *no current tasks exist.*
4. Current PM tasks exist where *no RCM-based tasks are recommended,* and are therefore candidates for deletion.

These comparison categories can be further refined to produce some charts with excellent visibility for management consumption, and examples of how this might be done are shown in Chap. 7, Figs. 7.5 to 7.9 inclusive. Further, the analyst can use the fourth category as a checklist to see if any obvious failure mode or PM task was inadvertently missed in the systems analysis process.

You will recall that in Sec. 5.2, it was recommended that the analyst defer the gathering of current PM task data during the information collection part of step 1 until this point in step 7 was reached. Hopefully,

RCM—Systems Analysis Process				
Step 7: Task selection				
Information: Comparison—RCM vs. current PM tasks		Rev no.:	Date:	
Plant:		Plant ID:		
System name:		System ID:		
Analysts:				
Component and failure mode	RCM-based task description	Freq.	Current task description	Freq.

Figure 5.20 Task comparison form.

you can now appreciate the basis for this recommendation; namely, to avoid biasing any portion of the RCM process with existing practices so that the comparison process done here could truly be viewed as two independent paths for defining a PM program for the same system.

Sanity check.   At key points throughout the systems analysis process, we have been collecting components and failure modes on a RTF list:

- On the SWBS, instruments were for "status information only."
- In the FMEA, their effects were local only.
- In the LTA, they were prioritized as Bin C or D/C.
- In the task selection process, no applicable task could be identified, or if the task was applicable, it was not considered to be effective in comparison to RTF.

The form on which we could collect such a list is shown in Fig. 5.21, which is now used to complete the sanity checklist process. The basis for such a sanity check derives from the possibility that there are valid reasons for performing a PM task even though the failure mode is not directly or solely related to a high-priority function. Referring to Fig. 5.21, we see seven such reasons listed on the form, and each individual situation may have other reasons which must be considered. The seven so listed are as follows:

1. *Marginal effectiveness.*   It is not totally clear that the RTF costs are significantly less than the PM costs.

2. *High-cost failure.*   While there is no loss of a critical function, the failure mode is likely to cause such extensive and costly damage to the component that it should be avoided.

3. *Secondary damage.*   Similar to item 2, except that there is a high probability that the failure mode could lead to extensive damage in *neighboring* components, and possibly loss of critical functions due to the domino effect.

4. *OEM conflict.*   The original equipment manufacturer recommends a PM task that is not supported by the RCM process. This dichotomy is especially sensitive if warranty conditions are involved.

5. *Internal conflict.*   Maintenance or operations personnel feel strongly about a PM task that is not supported by the RCM process. While these feelings can be more emotional than technical in their basis, management may decide against the RCM finding.

6. *Regulatory conflict.*   Stipulations by a regulatory body (e.g., Nuclear Regulatory Commission, EPA) have established a PM task

RCM—Systems Analysis Process										
Step 7: Task selection										
Information: Sanity checklist					Rev no.:			Date:		
Plant:					Plant ID:					
System name:					System ID:					
Analysts:										

Component and failure mode	Marginal effectiveness	High cost failure	Secondary damage	OEM conflict	Internal conflict	Regulatory conflict	Insurance conflict			RTF decision

Figure 5.21 Sanity checklist.

that is not supported by the RCM process. Should the RCM finding be argued with the regulators?

7. *Insurance conflict.* Similar to preceding items 4 and 6, and thus following the RCM finding would necessitate a change in the agreement with the insurance company.

A simple *yes* or *no* in each column will suffice to record the results of each consideration. Frequently, with a *yes* answer, there will be additional documentation to explain the nature of the conflict more thoroughly, and thus provide some basis for the ultimate decision on whether to approve or disapprove the RTF finding.

When the sanity check has been completed, the systems analysis process is essentially completed by updating the information on Figs. 5.19 and 5.20 to reflect any changes that occurred in the sanity checklist review. At this point in the process, Fig. 5.20 represents the summary listing of the RCM findings for the system in question. Note that most components will have more than one failure mode identified throughout the course of the FMEA exercise. Thus, the analyst must be careful to recognize that *all* failure modes must result in the RTF decision before the component itself can be declared as RTF. This caution, of course, does not deter the analyst from deciding that some component failure modes are RTF, while others require some PM action.

Each organization has its own culture with respect to a management review and approval process. As a general rule, Figs. 5.20 and 5.21 represent very good summary-level information for this purpose. At a minimum, the plant or facility manager and the maintenance and operations supervisor on his or her staff should be required to approve these results before implementation is initiated. A book containing the completed systems analysis information (i.e., the forms developed for steps 1 to 7) should be available as backup in order to show the details of the "how and why" behind each specific finding. Challenges to the findings are to be expected, and if the analyst has done the necessary homework, the conclusions and findings should be self-evident. But the process is not perfect, and adjustments during the management review are a constructive part of any RCM program.

5.9   Using Quantitative Reliability Data

You may have noticed by now that we have *not* used any quantitative reliability data in the RCM systems analysis process. In particular, we have not directly introduced any quantitative failure rate (λ) or reliability modeling data anywhere in the seven-step evaluation or prioritizing process. This is a very deliberate decision for the following reasons:

1. The ultimate decisions on PM task need and selection occur at the *failure mode level*. With the current data reporting systems at operating plants and facilities, there is rarely any credible quantitative reliability data collected at the failure mode level; what quantitative data is collected is found at the component level where PM task selections are not made (or should not be made). Thus, usable quantitative reliability history (for example, failure rate) is usually lacking where it might be helpful to the RCM process. This could change in the future, and perhaps should be reconsidered if such occurs.

2. In fact, however, there is no pressing need to introduce quantitative reliability data into the RCM systems analysis process. Realistic evaluations and decisions, from a maintenance point of view, can be made from the qualitative engineering and logic tree information that is systematically developed in the systems analysis process.

3. In addition, without quantitative data, the credibility of the results cannot be questioned on some abstract discussion of "numbers" validity. Only engineering know-how and related judgments are subject to challenge, and these areas can be more readily resolved.

4. Many people simply do not understand quantitative reliability values; thus their absence avoids unnecessary confusion and misunderstanding. (For example, did you read and understand App. B?)

While some RCM practitioners may feel differently about the preceding points, it is the author's experience that any introduction of quantitative reliability data or models into the RCM process only clouds the PM issue and raises credibility questions that are of no constructive value. Quantitative reliability data is not required in the selection of functions, the conduct of the FMEA, or the ordering of priorities in the LTA. It is useful in determining the validity of TD tasks and setting task intervals if the complete age-reliability relationship is known. In the majority of cases, however, the age-reliability relationship is not known with any degree of precision, even at the component level.

5.10   Information Traceability and Coding

It is a practical administrative consideration to address the question of information traceability. When RCM is applied to several systems in a plant, we find that the systems analysis information from steps 4 to 7 tends to pyramid, with the apex representing the system level of definition. Couple this with the possibility that several systems (i.e., pyramids) will eventually become the plant RCM program, and we can rather easily visualize the necessity for some accounting structure for the RCM information. Such an accounting structure will permit not only traceability down through a specific pyramid (i.e., system), but will also develop the structure that leads to the creation of an electronic file (if hard copy reports are not desired) and a computerized database of certain key data for future reference.

There are several ways to establish information coding for an accounting structure. In your particular situation, there may already be an active maintenance management information system which contains coding for the plants in your company, the systems in these plants, and the components in the systems. The primary need for coding, then, resides with the information that is peculiar to the RCM process. A simple way to handle this coding is shown below for a given system of interest:

```
Functional subsystem:   X
Function             :      .XX
Functional failure   :           .XX
Component            :                .XX
Failure mode         :                     .XX
Failure cause        :                          .XX
PM task              :                               .XX
```

Thus, for a given plant and system, each piece of RCM information will have a unique 13-digit number for identification and traceability purposes. While this may seem a bit cumbersome at first glance, this is actually not the case when the systems analysis information is committed to a computer for storage and processing. Furthermore, the value of such an accounting system becomes clearly evident, even with a single complex system, when you find it necessary to retrieve or cross-reference a piece of systems analysis data. With multiple systems, the numbering structure avoids what otherwise might well become an accounting quagmire by providing a unique identification and label for each piece of RCM data.

6

Illustrating RCM—
A Simple Example
(Swimming Pool Maintenance)

The best way to illustrate any analysis process is by way of example. Thus, we will devote this chapter to just such an endeavor, and will illustrate the RCM systems analysis process with an application to a home swimming pool (in this case, the pool in the author's backyard).

It is instructive to note that when I first acquired a home swimming pool, my knowledge of how to maintain it was zero. So I applied all of the current ad hoc PM methods (see Chap. 2, Sec. 2.5) to formulate my program—that is, experience (of which I had none), judgment (of which I had much, being an engineer), vendors' and friends' recommendations (which I later learned were based on experience and judgment), and a bit of brute force (if the filter could be dismantled, that must be the correct thing to do frequently). That was in 1975. By 1985, I was beginning to see the light! My early dealings with RCM prompted me to (slowly, but surely) change my PM style. Today, I have a 22-year-old pool that looks like new; for the 17 years that I have owned it, the water has never been removed and over 90 percent of the original components are intact and working like a charm. I hope they continue to do so!

In Chap. 6, we will follow the Seven-Step systems analysis process that was described in Chap. 5. But please recognize that some of the steps are rather easy to complete in comparison to what you would likely encounter in a more complex facility. Nonetheless, the principles are the same, and the illustration should help you to understand both the process and the mechanics of its implementation.

6.1 Step 1—System Selection
and Information Collection

System selection. The typical home swimming pool can be conveniently viewed as consisting of four major systems:

1. The *pool system* proper is where the fun takes place.

2. The *spa system* adjoins or is adjacent to the pool system. Not all pools will have this particular feature, but in this case there is an adjoining or attached spa enclosure that is integral to the pool proper. That is where relaxation takes place.

3. The *water treatment system* can be most easily identified at this point in our discussion as the group of equipment usually hidden in a corner someplace. That is where we keep the water "the way it should be."

4. The *utility system* supplies the electricity, gas, and water for the pool and its supporting equipment.

Steps 2, 3, and 4 will further describe these systems.

The system selection process in this example is quite easy because the only system with any significant equipment diversity (thus PM diversity) is the water treatment system. It can also be said from a qualitative point of view that the PM and CM costs are usually concentrated in the water treatment system. Thus, selection criteria 3 in Chap. 5, Sec. 5.2 is essentially the reason why we would want to initially address the *water treatment system*.

Information collection. As far as available information is concerned, this was not a lengthy task to perform because there wasn't much documentation available in the first place. In fact, the only information passed to me as the second owner of the pool was a handwritten set of instructions on how to align the valves in order to heat and use the spa. The previous owner did give me a walkdown which consisted of his collective O&M knowledge from five years of experience. The walkdown lasted all of 10 minutes, and my only recollection today of its content was the impression that if I threw some muriatic acid into the pool every now and then, all else would take care of itself. (As it turned out, this was *the* point to remember because, with hard water and rainfall as the water makeup source for evaporation, maintaining a neutral pH in the pool water is a key factor in avoiding any water problems.) Thus, for this illustration, I had to re-create the system schematic and virtually all of the information that is needed to perform the RCM systems analysis.

At this point, it would not surprise me if some of you who are reading this book aren't sitting there with a little smile on your face. Why? Because this scenario doesn't sound all that different from what you have experienced with your plant or facility—especially if it has been there for some time. Generally, in an older facility or plant (say, 10 years or older), basic information on the system P&ID, system design descriptions, O&M manuals, and most OEM component manuals (even part lists) may not be readily available. Fortunately, in most situations, there are plant personnel on site who have the essential elements of this data stored either in their desks or their minds. Also, OEM representatives stand ready to supply some information. Unfortunately, in this example, there was no prior experience still available to tap, and the OEM representatives came in the form of my local pool supply store, and in some instances, my neighbors. Suffice it to say, as noted earlier, information collection may be one of the more difficult tasks to accomplish in the whole systems analysis process, especially when dealing with older facilities that are more complex than a swimming pool.

Two essential pieces of information that were re-created in order to guide both the system selection and Steps 2 and 3 of the process were the pool block diagram and a pool schematic shown in Figs. 6.1 and 6.2, respectively. All remaining information was developed, as needed, for the subsequent Steps in the process.

6.2 Step 2—System Boundary Definition

Boundary definition for the water treatment system will follow the format and content described in Chap. 5, Figs. 5.4 and 5.5. The system schematic in Fig. 6.2 was used to specify the system boundary.

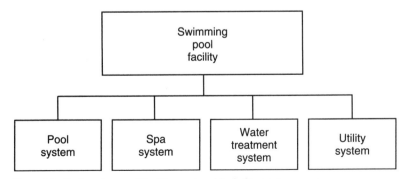

Figure 6.1 Swimming pool facility system block diagram.

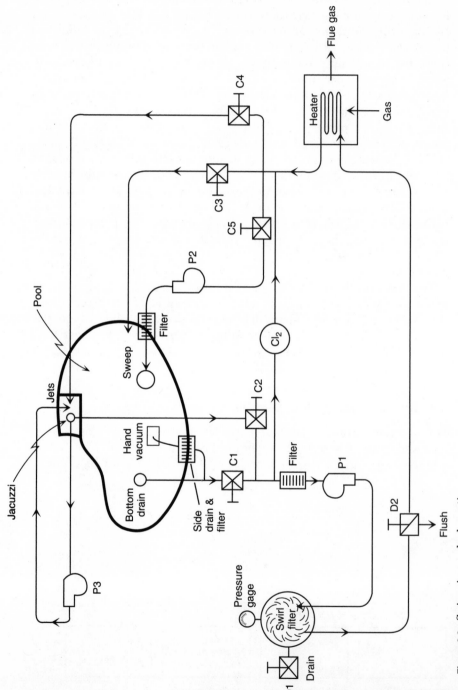

Figure 6.2 Swimming pool schematic.

Boundary overview

1. Major equipment included:

    Pumps and motors

    Heat exchanger

    Filters

    Valves and piping

    Chlorinator

    Various instruments

2. Primary physical boundaries:

    Water exits from the pool and spa

    Water entrances to the pool and spa

    Natural gas entrance to the heater and flue gas exit from the heater

    Electricity exiting the circuit breaker box

3. Caveats of note: None.

Boundary details

Type	Boundary system	Interface location
IN	Pool	Main water drain at bottom of pool
IN	Pool	Water skimmer at side of pool
OUT	Pool	Return water inlets (two) at side of pool
IN	Spa	Main water drain at bottom of spa
OUT	Spa	Return water inlet at side of spa
OUT	Sewer/drainage	Exit water drain on main filter
IN	Utility	Inlet side of gas valve on heater
OUT	Atmosphere	Exit gas ducts/openings on heat exchanger
IN	Utility	Pool side of electrical circuit breaker on 120 V supply
OUT	Pool	Quick disconnect on water line to the pool sweep
OUT	Sewer/drainage	Exit water flush line on main filter

6.3   Step 3—System Description and Functional Block Diagram

System description.   In this example, we can expand upon the pool system block diagram in Fig. 6.1 to define the functional block diagram for the water treatment system as shown in Fig. 6.3. This shows that we can conveniently divide the selected system into three functional subsystems—pumping, heating, and water conditioning. Since this is a fairly simple system, we will not divide the system description into the three subsystems, but rather will address the water treatment system as one entity.

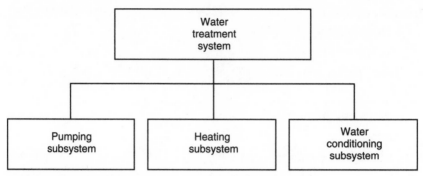

Figure 6.3   Functional block diagram for water treatment system.

Swimming pool facility description. As an introduction to the water treatment system description we will first discuss some features of the overall swimming facility. The facility, shown in Fig. 6.4, is a 34,000-gallon kidney-shaped pool that is typical of residential pool installations. In northern California (our location), it can be used about six months each year at a water temperature of 70°F or higher without any artificial heating—or year-round with artificial heating. The latter, without an enclosure, can be prohibitively costly, and virtually no one will heat a pool for year-round use. Pools are usually located for afternoon exposure to the sun, and in summer months the

Figure 6.4   Pool and spa system.

water temperature usually peaks out at about 85°F without any artificial heating.

The *pool system* has three functional subsystems:

1. *Water fill.*   This is simply the replacement of evaporated water with a garden hose (or, in some cases, with built-in water fill lines) to maintain a specified water level. Some pools are covered with a lightweight mylar blanket when not in use in order to minimize evaporation. The pool in our example here does not use any cover since it tends to be more of a nuisance than any real help if the pool is used frequently in warm weather.

2. *Manual water treatment.*   This consists of netting large debris (such as leaves), vacuuming the pool as needed, and adding muriatic acid to the water for pH control and an oxidizer to retain chlorine activity. The water chemistry is also checked periodically for pH level and chlorine content—at least weekly in hot weather and monthly in cold weather to guide the need for chemical additions.

3. *Pool sweep.*   As the name suggests, this is a water-driven device which sweeps about the pool with two subsurface water lines which operate in continuous motion, stirring the water to keep pool dirt in suspension for filtering. In recent years, some pools have installed automatic vacuum sweeps which continuously move about the pool sides and bottom with a suction action.

Many pools also have the *spa system,* and this is functionally, and often physically, tied to the main pool system. In this case, the spa system is a $6 \times 3 \times 3$ foot rectangular enclosure with two walls in common with the main pool. The water is usually heated in the 90 to 120°F range, depending on personal preferences, and is a very enjoyable form of relaxation at the end of "one-of-those-days." The spa system has two functional subsystems:

1. *Manual clean.*   This is the same as the preceding for debris removal and vacuuming.

2. *Water jets.*   Those of you familiar with spas know that high-velocity streams of water are pumped through "jets" in side locations to maintain a water circulation and swirl. In this spa system, there are three such jets. (The pressure behind these jets is not a part of the water treatment system.)

The *utility system* consists of the water, electricity, and gas services that supply the swimming pool facility.

Water treatment system (WTS).   The water treatment system is responsible for the majority of the functions required to maintain water purification. It also provides for artificial heating of the pool and spa water. These functions are achieved via three functional subsystems—namely, pumping, heating, and water conditioning. Our description here will follow the format of Fig. 5.7 in Chap. 5.

1. Functional description/key parameters

Pumping: The pumping subsystem provides two primary functions. First, it maintains a water flow of about 70 GPM circulating from the pool system through the heater and water conditioning subsystems. This flow includes a bleed through the line supplying water to the pool sweep in order to maintain a continuous priming flow for the pump in this line. Thus, the pool sweep cannot be operated unless the main water flow is in operation. Second, it provides the water flow and boost pressure, on demand, for the operation of the pool sweep. The pumping subsystem operates about 5 hours each day in the warm and hot seasons, and about 3 hours each day in the cooler months. These periods of operation are accomplished via automatic electromechanical switches that can turn both the main flow and pool sweep flow on and off at preset times. The pumping subsystem must be maintained in an airtight condition on the suction side of the water lines to preclude a loss of flow to the pumps. Also, during heavy rains in the winter months, the pumping subsystem must be able to drain water from the pool to avoid pool overflow.

Heating: The heating subsystem provides the capability to raise the ambient temperature of either the pool or spa water. Water circulation from the pumping subsystem continuously flows through a heat exchanger which is operated on natural gas, and when ignited, has an output rating of 383,000 Btu/hour. The heat exchanger is automatically controlled to provide the desired temperature to the pool (about 80°F) or the spa (90 to 120°F). It cannot efficiently heat both the pool and spa simultaneously. Since, by choice, virtually no heating of the pool occurs, the pool temperature control is simply maintained at a setting that is well below the ambient water temperature. The control unit also has a "hi-limit" temperature switch which will stop operation if exceeded. This high limit is set at about 140°F. Ignition is via a gas pilot flame which is shut off during the cooler months when neither the pool nor spa are used. The heating subsystem must be operated in a safe manner—that is, there must be no potential for personnel injury or death, or equipment damage from an uncontrolled fire or explosion.

Water conditioning: The water conditioning subsystem provides continuous automatic water filtering and chlorination treatment. Its function, then, is to maintain the water in a crystal clear condition. As noted previously, the filtering is augmented by periodic manual netting and vacuuming of the pool system, and by periodic manual addition of muriatic acid and oxidizer to the pool system. But the mainstay of the water-conditioning process is the daily operation of the pumping subsystem, which maintains the flow through the filter equipment and the chlorinator. Coarse filtering occurs at two locations: first, through a filter basket at the weir/skimmer water exit at the side of the pool (most of the water exit flow occurs here, not at the bottom drain in the pool), and second, through a filter basket immediately ahead of the main pump suction. Fine filtering is accomplished via a 70-gallon capacity swirl filter which uses diatomaceous earth as a filter medium to remove dust and particles not stopped by the basket filters. The swirl filter (so called because the internal design includes a series of semicircular plastic sections that rotate the water flow across their surfaces which are coated with the diatomaceous earth) is located on the discharge side of the main pump, and contains both a drain valve for use in removing excess water from the pool and a flush valve on its exit piping that can be used to periodically back-flush the swirl filter. A pressure gage on the swirl filter is used to calibrate the need for backflushing. The automatic chlorinator is simply a dispenser containing 1-inch chlorine tablets with a bleed line connected between the suction side piping and the heater exit piping. The bleed flow through this line can be adjusted up to 0.5 GPM to maintain a desired chlorine level in the pool water.

2. Redundancy features

With the exception of the two in-line basket filters, the water treatment system has no redundancy features.

3. Protection features

There are two important protection features. First, any electrical malfunction of consequence in the motors or instrumentation will trip the circuit breaker, thereby preventing catastrophic damage or fire. Second, the high-limit temperature control will prevent inadvertent excessive heating of the pool or spa water (and unnecessary gas consumption) should the desired temperature set point malfunction. There is also a grace period of 3–7 days (depending on water temperature) for the maintenance of acceptable water quality should any failure cause a complete shutdown of the water treatment system. This is accomplished by manual additions of liquid chlorine and other fungus-retarding chemicals.

4. Key instrumentation

The key instrumentation features of this subsystem are the electromechanical timers which provide for automatic operation of the pumping and water conditioning subsystems, and the high-limit temperature switch in the heating subsystem which was described previously. A manual switch is used to select heating for either the pool or spa, and a pressure gage is used to indicate flow status and clogging in the swirl filter.

Functional block diagram   In Fig. 6.3, we saw that the water treatment system had been divided into the three functional subsystems: pumping, heating, and water conditioning. In Fig. 6.5, we now complete the functional block diagram by also including the IN and OUT interfaces as well as the crucial interconnecting interfaces.

IN/OUT interfaces   Using the format shown in Fig. 5.9 in Chap. 5, we will now list all of the appropriate interfaces. Keep in mind that the RCM process assumes that IN interfaces are available when needed, and therefore we will concentrate later on using the OUT interfaces to identify and focus on function preservation and, ultimately, the selection of the PM tasks.

IN interfaces

- Water—from both the pool and spa
- AC power—120 V from house circuit breaker
- Natural gas—from house main supply line
- Chlorine—1-inch tabs and diatomaceous earth

OUT interfaces

- Treated water—to both pool and spa (treated = filtered, chlorinated, and heated, if desired)
- Signal status—timer, pressure, temperatures
- Flue gas
- Water—from overflow drain (upon demand)
- Water—from filter backflush (upon demand)
- Water—under pressure to pool sweep line

Internal OUT interfaces

- Water—from pump discharge to water conditioning at predetermined times
- Water—from swirl filter to heater in filtered condition
- Water—bleed flow from pump suction to Cl_2 canister
- Water—bleed flow from Cl_2 canister to heater exit pipe

6.3.4   System work breakdown structure (SWBS).   In the SWBS, we list the specific components that are associated with each of the three functional subsystems. The SWBS for the water treatment system is shown in Fig. 6.6.

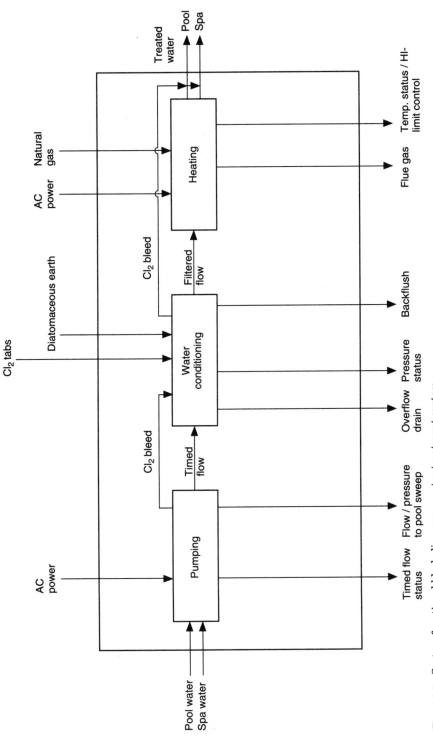

Figure 6.5 System functional block diagram—water treatment system.

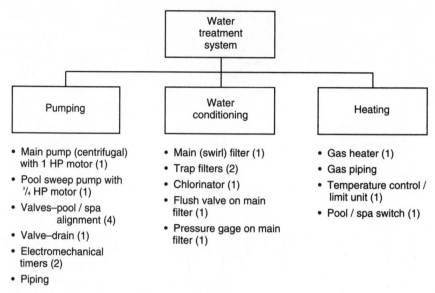

Figure 6.6 System work breakdown structure.

6.3.5 Equipment history. Our objective here is to recall (there are no formal work orders) the *corrective* maintenance (CM) actions that have occurred in the water treatment system. We will use this data, as applicable, in Step 5 when constructing the failure mode information. The equipment history information, as reconstructed from repair bills and personal repair experience, is shown in Fig. 6.7. As a point of interest, notice that the water treatment system operated virtually free of any unexpected problems (i.e., corrective maintenance actions) during the first 5–7 years of operation during my ownership (the equipment was 10–12 years old). Even then, most of the corrective maintenance problems (items 1, 5, 6, 7, 8) could have been avoided with timely and correctly applied preventive maintenance!

6.4 Step 4—System Functions and Functional Failures

We will now use the information developed in the system descriptions, IN/OUT interfaces and functional block diagram to formulate the specific function and functional failure statements. All of our effort to this point has been directed toward the ability to accurately list functions and functional failures in order to properly guide the eventual selection of the PM tasks.

The information so structured is shown in Fig. 6.8, and it follows the format of Fig. 5.12 in Chap. 5. Notice that we have started a number-

Component	Date	Failure mode	Failure cause
1. Alignment valves	1980 1988	Stuck (in closed/ open position)	Corrosion on stem during winter months
2. Pinhole leaks in suction piping	1981 1987 1991	Connecting joint deterioration	Aging
3. Main pump	1988	Bearing (sealed) breakdown	Aging
4. Main (swirl) filter-top canister	1986	Lip fracture (at joint with bottom canister)	Material flaw (mfr. replaced with no charge)
5. Main (swirl) filter—C clamp (top to bottom canister)	1982	Overstressed	Human error— excessive tightening (Oops!)
6. Flush valve on main filter	1984	Stuck closed	Lack of lubrication
7. Main filter pressure gage	1984	Erratic reading	Seal leak to atmosphere
8. Gas heater	1989	Erratic burner ignition	Contamination, corrosion

Figure 6.7 Equipment history—water treatment system.

ing system for this information which will be used in succeeding steps
to maintain traceability.

6.5   Step 5—Failure Mode and Effects Analysis

The initial action in Step 5 is to complete the equipment–functional
failure matrix shown in Chap. 5, Fig. 5.13. We do this by combining the
SWBS listings in Fig. 6.6 with the functional failure information in
Fig. 6.8 to produce the matrix shown in Fig. 6.9. This matrix, then,
becomes the road map to guide us in the FMEA, and is the "connecting
tissue" between the functions and equipment. As the matrix shows,
each item of equipment is associated with at least one functional failure, and 60 percent of the listed components are involved in two or
more functional failures. Thus, it would be premature for us to judge
that any of these components could be labeled *noncritical* and discarded from further consideration at this point in time (see Chap. 5,
Sec. 5.6, for some background discussion on this comment).

The FMEA is shown in Fig. 6.10, and the information presented here
is at the heart of the RCM process because it now identifies the specific
failure modes that can potentially defeat our functions as delineated in

1.0 Pumping Subsystem		
Function no.	Functional failure no.	Description
1.1		Maintain 70-GPM water flow at specified times to other subsystems.
	1.1.1	Fails to initiate flow at specified time.
	1.1.2	Flow is less than 70 GPM.
	1.1.3	Fails to terminate flow at specified time.
1.2		Maintain 50-GPM water flow at specified times to pool sweep line.
	1.2.1	Fails to initiate flow at specified time.
	1.2.2	Flow is less than 50 GPM.
	1.2.3	Fails to terminate before main flow shutdown.
1.3		Maintain water bleed to chlorinator.
	1.3.1	No bleed water flow.
1.4		Automatically activate/deactivate water flow.
	1.4.1	"On" and/or "Off" signals malfunction.
2.0 Water Conditioning Subsystem		
2.1		Provide filtered water to the heating subsystem.
	2.1.1	Fails to catch larger debris.
	2.1.2	Poor filtering efficiency (can be related to FF no. 1.1.2 above).
2.2		Send chlorinated water to exit piping.
	2.2.1	Fails to add chlorine to bleed water.
	2.2.2	No bleed water flow.
3.0 Heating Subsystem		
3.1		Provide desired heat input to water, on demand, at 383,000 Btu/hour.
	3.1.1	Fails to ignite.
	3.1.2	Fails to shut down at desired temperature.
3.2		Maintain a safe operation.
	3.2.1	Uneven burn and gas accumulation.
	3.2.2	Fails to shut down at Hi-limit control temperature.
	3.2.3	Full/partial stoppage of flue gas release.

Figure 6.8   Functions and functional failures.

Fig. 6.8. There are 38 unique failure modes listed, and they lead to 34 cases for carryover to the logic tree analysis (LTA). Due to the lack of any significant redundancy in the water treatment system, only four cases are dropped at this point in the analysis (and two of these because the failure mode in question is considered implausible).

6.6 Step 6—Logic (Decision) Tree Analysis

We can use the LTA structure shown in Chap. 5, Fig. 5.15 for the LTA in this swimming pool illustration. In this case, the "operators" are yours truly and his beloved better-half, and *the outage condition associated with question 3 in Fig. 5.15 will be defined as the inability to use either the pool or the spa.* As we have seen in the FMEA in Fig. 6.10, a majority of the "plant" effects are "pool/spa water deterioration" which, in their own right, are not the immediate initiators of outages. So we will see what happens to these failure modes as we progress through the LTA and then the sanity check in the final step of the systems analysis process.

Also notice that we have now established a simple coding system for identification and traceability per the discussion in Chap. 5, Sec. 5.10. For example, the first failure mode in Fig. 6.10 involving a "failed bearing on the main pump and 1-HP motor" could be coded for a computer entry as follows:

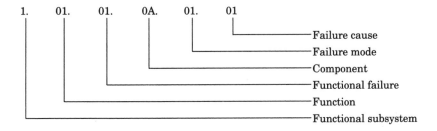

In this illustration, it is not really necessary to carry this coding system through in its entirety since we will not be committing the systems analysis information to a computer database. But our example does clearly show how a coding scheme can be easily employed when such is desired.

The LTA information is shown in Fig. 6.11, where 34 of the failure modes from the FMEA were carried over to the logic tree process. In summary, the LTA revealed the following categories (hence priorities):

$$A \text{ or } D/A = 2$$

$$B \text{ or } D/B = 8$$

$$C \text{ or } D/C = 24$$

As suggested in Chap. 5, Sec. 5.7, our initial action will be to relegate the 24 category C or D/C failure modes to the RTF status, and to pass the remaining 10 failure modes on to Step 7—task selection.

			Pumping							
Equipment		Functional failures	1.1.1 Fails to initial flow at specified time (pool/spa)	1.1.2 Flow <70 GPM	1.1.3 Fails to terminate flow at specified time	1.2.1 Fails to initiate (pool sweep)	1.2.2 Flow <50 GPM	1.2.3 Fails to terminate before main flow stops	1.3.1 No bleed water flow	1.4.1 On/Off signals malfunction
No.	Name									
A	Main pump with 1-HP motor		X	X						
B	Pool sweep pump with ¾-HP motor					X	X			
C	Valves—pool/spa alignment (2)		X							
D	Valve—drain		X							
E	Electromechanical timers (2)		(1)		(1)	(1)		(1)		X
F	Water piping		X			X	X		X	
G	Main (swirl) filter			(2)						
H	Trap filters (2)			(2)						
I	Chlorinator									
J	Flush valve on main filter									
K	Pressure gage									
L	Gas heater									
M	Gas piping									
N	Temperature control/limit unit									
O	Pool/spa switch									

(1) Covered in FF No. 1.4.1
(2) Covered in FF Nos. 2.1.1 and 2.1.2

Figure 6.9   Equipment–functional failure matrix.

	Water Conditioning				Heating				
	2.1.1 Fails to catch larger debris	2.1.2 Poor filtering efficiency	2.2.1 Fails to add CL_2 to bleed water	2.2.2 No bleed water flow	3.1.1 Fails to ignite	3.1.2 Fails to shutdown at desired temp.	3.2.1 Uneven burn and gas accumulation	3.2.2 Fails to shut down at Hi-limit control temp.	3.2.3 Full/partial stoppage of flue gas release
				X					
		X							
	X	X							
			X						
		X							
		X							
					X		X		X
					X		X		
					X	X		X	
					X				

Subsystem: Pumping

Functional Failure: 1.1.1—Fails to initiate flow at specified time (pool/spa)

Equipment	Failure mode	Failure cause	Failure effect Local	Failure effect system (pool/spa)	Plant	LTA
Main pump and 1-HP motor	.01—Failed bearing (sealed)	.01.01—Age/wearout	Pump inoperative	Loss of all flow	Pool/spa water deterioration—spa inoperative	Y
	.02—Motor short/ground	.02.01—Insulation aging	Pump inoperative	Loss of all flow	Pool/spa water deterioration—spa inoperative	Y
	.03—Leak (at pump motor joint)	.03.01—Broken gasket .03.02—Loose bolts	Loss of pump suction—Possible motor/pump damage	Loss of all flow	Pool/spa water deterioration—spa inoperative	Y
Alignment valves (4)	.01—Stuck in open or closed position	.01.01—Corrosion/ contamination	Cannot open (or close) valve on demand	Cannot realign flow for pool to spa or vice versa	Either pool or spa inoperative	Y
Drain valve	.01—Stuck in closed (normal) position	.01.01—Corrosion	Cannot open valve on demand	Cannot drain water from system	Pool/spa can overflow in heavy rains	Y
Timers (2) (main flow and pool sweep)	See FF no. 1.4.1					
Piping	.01—Rupture	.01.01—Material flaw	Considered to be implausible failure mode			N
	.02—Pinhole leak (incl. joints)	.02.01—Corrosion	Loss of pump suction (could temporarily restore)	Flow deterioration	Pool/spa water deterioration	Y

Functional Failure: 1.1.2—Flow <70 GPM

Main pump with 1-HP motor	.01—Bearing deterioration	.01.01—Age/wearout	Erratic pump operation	Reduced flow	Pool/spa water deterioration	Y
Main (swirl) filter	See FF no. 2.1.2					
Trap filters (2)	See FF no. 2.1.1					

Functional Failure: 1.1.3—Fails to terminate flow at specified time

Timers (2)	See FF no. 1.4.1

Functional Failure: 1.2.1—Fails to initiate flow at specified time (pool sweep)

Pool sweep pump with ¾-HP motor	.01—Failed bearing (sealed)	.01.01—Age/wearout	Pump inoperative	Loss of flow to pool sweep	Pool/spa water deterioration	Y
	.02—Motor short/ground	.02.01—Insulation aging	Pump inoperative	Loss of flow to pool sweep	Pool/spa water deterioration	Y
	.03—Leak (at pump motor joint)	.03.01—Broken gasket .03.02—Loose bolts	Loss of pump suction—possible pump/motor damage	Loss of flow to pool sweep	Pool/spa water deterioration	Y
Timers (2)	See FF no. 1.4.1					
Piping	Same as FF no. 1.1.1 above					

Figure 6.10 Failure mode and effects analysis.

Subsystem: Pumping

Equipment	Failure mode	Failure cause	Local	Failure effect system	Plant	LTA
Functional Failure: 1.2.2—Flow <50 GPM						
Pool sweep pump with ¾-HP motor	.01—Bearing deterioration	.01.01—Age/ wearout	Erratic pump operation	Reduced flow to pool sweep	Pool/spa water deterioration	Y
Piping	.01—Clogged filter (on pool sweep line)	.01.01—Debris buildup	Reduced filter efficiency	Reduced flow to pool sweep/over-worked pump	Pool/spa water deterioration	Y
Functional Failure: 1.2.3—Fails to terminate before main flow stops						
Timers (2)	See FF no. 1.4.1					
Functional Failure: 1.3.1—No bleed water flow (to chlorinator)						
Piping (⅜-inch neoprene bleed line)	.01—Rupture	.01.01—Age	Loss of pump suction (could temporarily restore)	Flow deterioration	Pool/spa water deterioration	Y
	.02—Pinhole leak (incl. joints)	.02.01—Age	Loss of pump suction (could temporarily restore)	Flow deterioration	Pool/spa water deterioration	Y
Functional Failure: 1.4.1—On/Off signals malfunction						
Timer—to main pump	.01—Failed clock	.01.01—Age/ wearout	Loss of automatic timing	System fails to start or system fails to stop	Pool/spa water deterioration / None	Y
	.02—Short circuit	.02.01—Insulation aging	Loss of automatic timing	System fails to start or system fails to stop	Pool/spa water deterioration / None	Y
	.03—Set points (mechanical) loose	.03.01—Wear/vib. .03.02—Improperly installed	Loss of automatic timing	System fails to start or system fails to stop	Pool/spa water deterioration / None	Y

Timer—to pool sweep pump	Same as above	Same as above	Loss of automatic timing *Note:* Timer can be manually operated *if* someone observes failure state	Pool sweep fails to start or pool sweep fails to stop and motor can burn out	Pool/spa water deterioration	

Subsystem: Water Conditioning

Functional Failure: 2.1.1—Fails to catch larger debris

Trap filter (at Weir)	.01—Broken basket	.01.01—Mishandled .01.02—Age	Hole in plastic basket	Large debris escapes to second trap filter	None	Y
	.02—Clogged	.02.01—Debris buildup	Reduced filter efficiency	Reduced flow/over-worked pump	Pool/spa water deterioration	Y
Trap filter (at main pump suction)	.01—Broken basket	.01.01—Mishandled .01.02—Age	Hole in plastic basket	Large debris escapes to swirl filter	None	Y
	.02—Clogged	.02.01—Debris buildup	Reduced filter efficiency	Reduced flow/over-worked pump	Pool/spa water deterioration	Y
	.03—Leaky gasket (on filter cover)	.03.01—Age	Loss of pump suction (could temporarily restore)	Flow deterioration	Pool/spa water deterioration	Y

Functional Failure: 2.1.2—Poor filtering efficiency

Trap filters (2)	Same as FF no. 2.1.1 above					
Main (swirl) filter	.01—Clogged	.01.01—Debris and dirt buildup	Reduced filter efficiency	Reduced flow/over-worked pump	Pool/spa water deterioration	Y

Figure 6.10 (*Continued*)

Subsystem: Water Conditioning

Equipment	Failure mode	Failure cause	Failure effect Local	system	Plant	LTA
		Functional Failure: 2.1.2—Poor filtering efficiency (continued)				
	.02—Water leak (at top to bottom section joint)	.02.01—Aging gasket	Water dripping from filter joint	None	None	N
Flush valve on main filter	.01—Stuck	.01.01—Corrosion	Inoperative back-flush valve	Cannot backflush main filter	None	Y
Pressure gage	.01—False reading (lower than actual)	.01.01—Age	Erroneous pressure signal	Reduced flow/over-worked pump (*Note:* Assumes main filter is clogged)	Pool/spa deterioration	Y
		Functional Failure: 2.2.1—Fails to add Cl_2 to bleed water				
Chlorinator	.01—Clogged	.01.01—Debris from undissolved tabs	No flow from chlorinator	No chlorine injection to bleed line	Pool/spa water deterioration	Y
	.01—No tabs	.01.01—Forgot to refill	Empty chlorinator	No chlorine injection to bleed line	Pool/spa water deterioration	Y
		Functional Failure: 2.2.2—No bleed water flow				
Piping (⅜-inch neoprene bleed line)	.01—Rupture	.01.01—Age	Broken exit line from chlorinator	No chlorine injection to pool return piping	Pool/spa water deterioration	Y
		Subsystem: Heating				
		Functional Failure: 3.1.1—Fails to ignite				
Gas heater	.01—Failed pilot light	.01.01—Wind or rain-storm	No pilot light and smell of gas	Heater will not ignite on demand	Spa cannot be used	Y
Gas piping	.01—Blockage (in heater piping)	.01.01—Large foreign object in line	Considered to be impossible failure mode			N

Component	Failure mode	Cause	Local effect	Effect on heater	Effect on system	Critical
Pool/spa switch	.01—Failed switch	.01.01—Aging	No electrical contact through switch	Gas valve will not open—heater will not ignite	Spa cannot be used	Y
Temperature control/limit unit	.01—Control unit fails (Lo)	.01.01—Random part failure	Control unit inoperative	Heater will not ignite on demand	Spa cannot be used	Y
Functional Failure: 3.1.2—Fails to shut down at desired temperature						
Temperature control/limit unit	.01—Control unit fails (Hi)	.01.01—Random part failure	Control unit inoperative	Heater will not automatically shut down	Eventual over-temperature in spa; unnecessary gas consumption	Y
Functional Failure: 3.2.1—Uneven burn and gas accumulation						
Gas heater	.01—Burner dirty/clogged	.01.01—Corrosion dirt, insects	Delay in simultaneous ignition across burner	Small to large explosion in heater, possibly fire and severe heater damage	Spa cannot be used	Y
Gas piping	.01—Leak (at connection)	.01.01—Age/vibration	Smell of gas	None	None	N
Functional Failure: 3.2.2—Fails to shut down at Hi-limit control temperature						
Temperature control/limit unit	Same as FF no. 3.1.2 above					
Functional Failure: 3.2.3—Full/partial stop of flue gas release						
Gas heater	.01—Clogged vents	.01.01—Leaves, pine needles, insects, etc.	Blocked vents	Flue gas cannot escape—possible fire and damage to heater	Spa cannot be used	Y

Figure 6.10 (*Continued*)

| Functional failure | Component/failure mode | Criticality analysis | | | Category | Comment |
		Evident?	Safety?	Outage?		
1.1.1—Fails to initiate flow at specified time	*Main pump and 1-HP motor*					
	.01—Failed bearing	Y	N	Y	B	Must be corrected in ≤4 days or *serious* water deterioration occurs.
	.02—Motor short	Y	N	Y	B	Ditto above.
	.03—Leak	Y	N	Y	B	Ditto above, plus can cause *serious motor/pump* damage if not shut off in ≤4 hours.
	Alignment valves					
	.01—Stuck open or closed	N	N	Y	D/B	
	Drain valve					
	.01—Stuck closed	N	N	Y	D/B	Spa cannot be used if pool water level is too high.
	Piping					
	.02—Pinhole leak	Y	N	N	C	Must be corrected in ≤4 days or *serious* water deterioration occurs, plus can cause *serious motor/pump* damage if not shut off in ≤4 hours.
1.1.2—Flow <70 GPM	*Main pump and 1-HP motor*					
	.01—Bearing deterioration	Y	N	N	C	Gives audible indication.
1.2.1—Fails to initiate flow at specified time (pool sweep)	*Pool sweep pump with ¾-HP motor*					
	.01—Failed bearing	Y	N	N	C	
	.02—Motor short	Y	N	N	C	
	.03—Leak	Y	N	N	C	Can cause *serious* motor/pump damage if not shut off ≤4 hours.

1.2.2—Flow <50 GPM (pool sweep)	*Pool sweep pump with ¾-HP motor*			
	.01—Bearing deterioration	Y	N	C	Gives audible indication.
	Piping .01—Clogged filter (on pool sweep line)	N	N	D/C	Can shorten motor life if clogged for several months.
1.3.1—No bleed water flow (to chlorinator)	*Piping (neoprene line)*			
	.01—Rupture	Y	N	C	Must be corrected in ≤4 days or *serious* water deterioration occurs, plus can cause serious motor/pump damage if not shut off in ≤4 hours.
	.02—Pinhole leak	Y	N	C	Ditto above.
1.4.1—On/Off signals malfunction	*Timer—to main pump*			
	.01—Failed clock	Y	N	C	Main concern is failure to start. Could lead to pool sweep motor damage when pool sweep timer initiates.
	.02—Short circuit	Y	N	C	Ditto above.
	.03—Set points loose	Y	N	C	Ditto above.
	Timer—to pool sweep				Same as timer—to main pump, except that main concern here is failure to stop. Could lead to pool sweep motor damage when main pump timer deactivates.

Figure 6.11 Logic tree analysis.

Functional failure	Component/failure mode	Criticality analysis				Comment
		Evident?	Safety?	Outage?	Category	
2.1.1—Fails to catch larger debris	*Trap filter (at Weir)*					
	.01—Broken basket	N	N	N	D/C	Increases debris buildup in second trap filter.
	.02—Clogged	N	N	N	D/C	Could shorten motor life if clogged for several weeks.
	Trap filter (at main pump suction)					
	.01—Broken basket	N	N	N	D/C	Increases debris buildup in main swirl filter.
	.02—Clogged	N	N	N	D/C	Could shorten motor life if clogged for several weeks.
	.03—Leaky gasket	Y	N	N	C	Must be corrected in ≤4 days or *serious* water deterioration occurs, plus can cause serious motor/pump damage if not shut off in ≤4 hours.
2.1.2—Poor filtering efficiency	*Main (swirl) filter*					
	.01—Clogged	Y	N	N	C	Pressure gage reading is increasing. Can shorten motor life if clogged for serveral weeks.
	Flush valve on main filter					
	.01—Stuck	N	N	N	C	
	Pressure gage					
	.01—False reading	N	N	N	D/C	Would remove capability to easily discern that filter clogging is occurring.
2.2.1—Fails to add Cl₂ to bleed water	*Chlorinator*					
	.01—Clogged	N	N	N	D/C	Must be corrected in ≤4 days or *serious* water deterioration occurs.
	.02—No tabs	N	N	N	D/C	Ditto above.

2.2.2—No bleed water flow	*Piping (neoprene line)*				
	.01—Rupture	N	N	N	D/C	Ditto above.
3.1.1—Fails to ignite	*Gas heater*				
	.01—Failed pilot light	Y	N	Y	B	Can smell gas.
	Pool / spa switch				
	.01—Failed switch	N	N	Y	D/B
	Temperature control / limit unit				
	.01—Control unit fails (Lo)	N	N	Y	D/B
3.1.2—Fails to shut down at desired temperature	*Temperature control / limit unit*				
	.01—Control unit fails (Hi)	Y	N	N	C
3.2.1—Uneven burn and gas accumulation	*Gas heater*				
	.01—Burner clogged/dirty	N	Y	Y	D/A
3.2.3—Full/ partial stoppage of flue gas release	*Gas heater*				
	.01—Clogged vents	Y	Y	Y	A

Figure 6.11 Logic tree analysis (*Continued*)

FF no.	Component/ failure mode	Failure cause	Selection guide 1	2	3	4	5	6	7	8	Candidate tasks	Effectiveness information	Sel. dec.	Est. freq.
1.1.1	*Main pump and 1-HP motor*													
	.01—Failed bearing	.01.01—Age/ wearout	N	N	N	N	—	N	N	N	.01—Vibration monitoring .02—RTF	.01 is not considered cost effective.	RTF	—
	.02—Motor short/ground	.02.01— Insulation aging	N	N	N	N	—	—	N	N	.01—RTF	There is no practical way to stop or detect onset of this failure mode.	RTF	—
	.03—Leak	.03.01—Broken gasket	N	N	Y	N	—	Y	—	—	.01—Inspect for signs of water seepage at the gasket area. (CD) .02—RTF	.01 is the most cost effective.	.01 CD	6 mo.
		.03.02—Loose bolts	N	N	Y	N	—	Y	—	—				
1.1.1	*Alignment valves*													
	.01—Stuck open or closed	.01.01— Corrosion/ contamination	P	Y	N	Y	Y	Y	—	—	.01—Check the valve operation in early spring (before use of spa). (FF) .02—Lubricate valve stem. (TD) .03—RTF	.01 is the most cost effective since history indicates infrequent sticking.	.01 FF	12 mo. (spring)
1.1.1	*Drain valve* .01—Stuck closed	.01.01— Corrosion	P	Y	N	Y	Y	Y	—	—	.01—Check valve operation in early fall (before rainy season). (FF)	.01 is the most cost effective since history indicates no sticking.	.01 FF	12 mo. (fall)

132

3.1.1	*Gas heater*								
	.01—Failed pilot light								
	.01.01—Wind or rainstorm	Y	N	N	—	—	—	N	.02—Exercise valve periodically and lubricate. (TD)	Do not choose RTF since sticking valve could lead to backyard flooding.	—
									.03—RTF		
									.01—RTF	There is no way to stop or detect onset of failed pilot light. Smell of gas near heater is very evident.	RTF
3.1.1	*Pool/spa switch*								
	.01—Failed switch								
	.01.01—Aging	N	N	Y	Y	N	—	N	.01—Periodically check switch for proper operation.	.02 is the most cost effective.	—
									.02—RTF	Switch has never failed. If it did, replacement is quick and cheap.	RTF
3.1.1	*Temp. control/ limit unit*								
	.01—Control unit fails (Hi)								
	.01.01—Random part failure	N	N	Y	N	N	—	N	.01—RTF There is no applicable PM task for the failure cause.	.01 is the only option. Unit has not failed to date.	RTF

Figure 6.12 Task selection.

FF no.	Component/ failure mode	Failure cause	Selection guide								Candidate tasks	Effectiveness information	Sel. dec.	Est. freq.
			1	2	3	4	5	6	7	8				
3.2.1	*Gas heater* .01—Burners clogged/dirty	.01.01—Corrosion, dirt, insects	Y	Y	Y	Y	N	Y	—	—	.01—Remove burner unit and clean—repair as required. (TD)	.01 appears to be the most cost effective.	.01 TD	60 mo.
											.02—Observe delay in ignition of all burners. (CD)	It is questionable if .02 is realistic. RTF is not an option here due to category A rating.		
											.03—RTF			
3.2.3	*Gas heater* .01—Clogged vents	.01.01—Leaves, pine needles, insects, etc.	Y	Y	N	N	—	Y	—	—	.01—Clean gas heater vents in early spring (before use of spa). (TD)	.01 is the most effective. RTF is not an option here due to category A rating.	.01 TD	12 mo. (spring)
											.02—RTF			

Figure 6.12 Task selection (*Continued*)

6.7 Step 7—Task Selection

Task selection process.   The category A, D/A, B, and D/B items from the LTA are put through the task selection process described in Chapter 5, Figs. 5.18 and 5.19. The results of this process are shown in Fig. 6.12. Of the ten top-priority failure modes derived from the LTA, we defined two TD tasks, one CD task, two FF tasks and five RTF decisions. The latter RTF decisions were driven in three cases by the fact that no *applicable* task could be identified, and in two cases by the *effectiveness* consideration (see Chap. 4, Sec. 4.4, for discussion of "Applicable and Effective").

Sanity check.   Figure 6.13 lists each component and its related failure mode that was assigned as a category C or D/C priority from the LTA. The listing also contains the two items that were dropped at the FMEA (i.e., the main swirl filter section leak and the gas piping leak). Of the

Component/ failure mode	Marginal effectiveness	High cost failure	Secondary damage	OEM conflict	Internal conflict	Regulatory conflict	Insurance conflict		RTF decision
Main pump and 1-HP motor .01—Bearing deterioration	X								No—CD task
		Listen for discernible increase in pump/motor noise level—6 months. (This was successfully used to predict need for motor/pump replacement in 1984 *before* total failure.)							
Piping .02—Pinhole leak									RTF—See note 3
Main (swirl) filter .01—Clogging	X								No—CD task
		Monitor filter pressure gage reading for pressure increase above approximately 20 psi—6 months.							
.02—Water leak (at tank sections joint)									RTF

Figure 6.13 Sanity check.

Component/ failure mode	Marginal effectiveness	High cost failure	Secondary damage	OEM conflict	Internal conflict	Regulatory conflict	Insurance conflict		RTF decision
Trap filters (both) .01—Clogging	X								No—TD task
Remove filter baskets, inspect and clean—3 months.									
.01—Broken basket	X								No— included in above
Same TD task as above. Replace basket if necessary.									
Trap filter (at main pump suction) .03—Leaky gasket	X								No—TD task
Replace gasket—24 months.									
Pool sweep pump with ¾-HP motor .01—Failed bearing									RTF—See note 1
.02—Motor short									RTF—See note 2
.03—Leak		X							No—CD task
Inspect for signs of water seepage at the gasket area—6 months.									
.01—Bearing deterioration	X								No—CD task
Listen for discernible increase in pump/motor noise level—6 months.									
Piping (on pool sweep line) .01—Clogged filter	X								No—TD task
Remove and clean filter—12 months.									

Figure 6.13 Sanity check (*Continued*)

Component/ failure mode	Marginal effectiveness	High cost failure	Secondary damage	OEM conflict	Internal conflict	Regulatory conflict	Insurance conflict		RTF decision
⅜-inch neoprene bleed line									
.01—Rupture									RTF—See note 3
.02—Pinhole leak									RTF—See note 3
Timers (both)									
.01—Failed clock									RTF—See note 3
.02—Short circuit									RTF—See note 3
.03—Loose set points	X		X						No—FF task
Check to assure that set point screws are tight—3 months.									
Flush valve on main filter .01—Stuck	X								No—TD task
Operate valve and lubricate if necessary—6 months.									
Pressure gage .01—False reading	X								No—CD task (see note 4)
When monitoring gage for pressure reading, assure that gage returns to zero when main pump is turned off—6 months.									
Chlorinator .01—Clogged	X								No—TD task
Remove and clean dispenser unit—12 months.									

Figure 6.13 Sanity check (*Continued*)

Component/ failure mode	Marginal effectiveness	High cost failure	Secondary damage	OEM conflict	Internal conflict	Regulatory conflict	Insurance conflict		RTF decision
.02—No tabs	X								No—TD task
			Refill Cl$_2$ tabs. 2 months—off season 2 weeks—during season						
Gas piping .01—Leak									RTF—See note 3
Temp. control / limit unit .01—Control unit fails (Hi)									RTF

Notes:

1. While some form of vibration monitoring may be possible, this is not considered to be a cost-effective approach. Motor and/or pump are easily replaceable if required, but expected life is 15 to 20 years.

2. A motor short, due to rain penetration or long-term deterioration, is impossible to prevent via any PM action, and onset is virtually impossible to detect.

3. It would be desirable to prevent or predict this failure mode, but there really is no practical way to do it.

4. This visual check does not absolutely guarantee that the pressure gage is reading correctly, but observing its performance for erratic behavior when the pump is turned on and off is a reasonable indicator of unreliable readings. Pressure gage should be replaced if performance is suspect.

Figure 6.13   Sanity check (*Continued*)

23 *unique* failure modes subjected to the sanity check, only 10 retained the RTF status; 13 items were assigned a PM task because of the marginal effectiveness or secondary damage potential associated with an RTF decision. In selecting the PM tasks here, the task selection process rationale in Fig. 6.12 was employed, but the formality of completing the form was not accomplished. This form is a matter of choice, and could be used if the analyst so elected.

Task comparison. Figure 6.14 presents the results of the task selection process and sanity check for each component in the water treatment system. The components are listed in the same order as shown in Fig. 6.9, and the results are shown by failure mode per component. The

Component/ failure mode	RCM-based task description	Freq.	Previous task description	Freq.
Main pump with 1-HP motor .01—Failed bearing	RTF		None	
.02—Motor short	RTF		None	
.03—Leak	Inspect for signs of water seepage at the gasket area.	6 months (CD)	None	
.01—Bearing deterioration	Listen for discernible increase in pump/motor noise level.	6 months (CD)	None. (The noise level detection was accidentally discovered in 1984.)	
Pool sweep pump with ¾-HP motor .01—Failed bearing	RTF		None	
.02—Motor short	RTF		None	
.03—Leak	Inspect for signs of water seepage at the gasket area.	6 months (CD)	None	
.01—Bearing deterioration	Listen for discernible increase in pump/motor noise level.	6 months (CD)	None	
Valves—pool / spa alignment .01—Stuck open or closed	Check the valve operation in early spring (before use of spa).	12 months (FF)	None (actually stuck twice—could not use spa until time was available to fix).	
Valve—drain .01—Stuck closed	Check valve operation in early fall (before rainy season).	12 months (FF)	None	
Electromechanical timers .01—Failed clock	RTF		None	
.02—Short circuit	RTF		None	

Figure 6.14 Task comparisons.

Component/ failure mode	RCM-based task description	Freq.	Previous task description	Freq.
Electromechanical timers (Cont.)				
.03—Loose set points	Check to assure that set-point screws are tight (especially before a vacation trip).	3 months (FF)	None	
Water piping				
.01—Rupture	RTF		None	
.02—Pinhole leak (includes neoprene bleed line)	RTF		None	
.01—Clogged filter (pool sweep line)	Remove and clean filter	12 months (TD)	None	
Main (swirl) filter				
.01—Clogging	Monitor filter pressure gage reading for pressure increase above approx. 20 psi.	6 months (CD)	Disassemble filter and clean.	6 months (TD)
.02—Water leak (at tank section joint)	RTF		None	
Trap filters				
.01—Clogging	Remove filter baskets, inspect, and clean.	3 months (TD)	Check filter baskets and clean after a storm.	Varied (TD)
.01—Broken basket	Same as above.		Same as above.	
.03—Leaky gasket (at main pump suction)	Replace gasket.	24 months (TD)	None	
Chlorinator				
.01—Clogged	Remove and clean dispenser unit.	12 months (TD)	None	
.01—No tabs	Refill Cl$_2$ tabs (more frequently during warm season).	2 months or 2 weeks (TD)	Refill Cl$_2$ tabs	2 months or 2 weeks (TD)

Figure 6.14 Task comparisons (*Continued*)

Component/ failure mode	RCM-based task description	Freq.	Previous task description	Freq.
Flush valve on main filter .01—Stuck	Operate valve and lubricate if necessary.	6 months (TD)	None	
Pressure gage .01—False reading	When monitoring gage for pressure reading, assure that gage returns to zero when main pump is turned off.	6 months (CD)	None	
Gas heater .01—Burners clogged/dirty	Remove burner unit and clean—repair as required.	60 months (TD)	None	
.01—Clogged vents	Clean gas heater vents in early spring (before use of spa).	12 months (TD)	Clean gas heater vents in early spring (before use of spa).	12 months (TD)
.01—Failed pilot light	RTF		None	
Gas piping .01—Leak	RTF		None	
Temperature control / limit unit .01—Control unit fails (Hi or Lo)	RTF		None	
Pool / spa switch .01—Failed switch	RTF		None	

Figure 6.14 *(Continued)*

point that stands out here is that the previous PM tasks were essentially representative of a reactive PM program. The large number of "none" entries (for 26 of the final list of 31 unique failure modes) says that maintenance was done, for the most part, by fixing things when they broke (i.e., corrective maintenance). In the previous PM program, there were no deliberate decisions to "RTF," although the "none" entries could be considered as "RTF by default."

The RCM process introduced a rational PM program to the swimming pool and, quite frankly, helped me to avoid a rash of bothersome (and sometimes harmful to the equipment) failure events. As one example, I can recall going on a one-week vacation only to return and find that the main pump was not shutting off; the set screw on the "off" switch came loose. I don't know how long it had been running, but I was lucky that it was not the "on" switch that came loose because this would have damaged (maybe burned out) the pool sweep motor! I was also very pleased to stop my biannual ritual of disassembling the main (swirl) filter for cleaning.

We can summarize the task comparisons, *by component,* as follows:

	RCM	Previous
TD	5	4
CD	4	0
FF	3	0
NONE	N/A	11
RTF	3	N/A

and

RCM = Previous (including components that are both RTF and none)	3
RCM = Previous, but modified	4
RCM, but no previous	8
Previous, but no RCM	0

In other words, the RCM process directed me to make deliberate PM decisions where such had never occurred, and to reexamine the validity of my time-directed tasks. The net result has been trouble-free (virtually no corrective maintenance) operation for over 7 years.

Illustrating RCM—Examples Drawn from Industry

The Seven-Step systems analysis process described in Chap. 5 has been successfully applied at several utility power generation plants. In this chapter, we will review some selected results from these real-world RCM programs.

Most of the RCM applications have been aimed at optimizing existing PM programs, and examples here in this category are drawn from RCM efforts at GPU Nuclear Corporation and Florida Power and Light Company. In one recent application, RCM was used as the methodology to define the PM program for a relatively new system in a combined-cycle plant that was designed, built, and is currently operated by Westinghouse Electric Corporation. This example is shown here to illustrate the ability to successfully apply RCM in an area where little prior experience existed—a fact that is not always recognized by some analysts who are new to RCM. Finally, some results from a pilot RCM project at a USAF engine test facility are included to indicate yet another area where RCM is being tested. This latter effort is conducted by Sverdrup Technology, Inc. for the USAF.

The results from these examples will be illustrated in four ways:

1. An overview of the systems analysis process and PM tasks in a summary statistical format

2. Selected in-depth examples of the RCM methodology using a "vertical slice" through the systems analysis process

3. Discussion of selected PM task results to illustrate how RCM can optimize a PM program

4. Discussion of peripheral benefits realized from RCM programs

7.1 Overview of RCM Results

The RCM systems analysis process described in Chap. 5 has been applied by the author at numerous plants and facilities during the last 10 years. In this chapter, we will take some examples from these applications and present selected extracts that will illustrate both the RCM process and typical results that have been realized from the process.

In this section, we will look at three different profiles which depict the results of the RCM application to six different facilities. The profiles are summaries of statistical data which can be extracted from the RCM analysis in order to illustrate top-level results and findings. These profiles have been used frequently by the author with much success in presentations to management where it is necessary to visibly display pertinent project information and results with some reasonable dispatch.

The plants and facilities that are used in the examples are as follows:

1. *GPU Nuclear Corporation, Three Mile Island nuclear power plant.* TMI-1, located near Harrisburg, Pennsylvania, is an 870 MW$_e$ nuclear power plant. It is a pressurized water reactor design that was supplied by Babcock and Wilcox. A simplified plant schematic is shown in Fig. 7.1. Our examples here are drawn from the RCM analyses performed on the main feedwater and instrument air systems. The TMI RCM program is perhaps the most comprehensive RCM effort in place today in the U.S. utility industry. It is a 5-year program with the objective of completing about 30 RCM systems analyses on the plant. (A typical nuclear plant has over 100 unique systems.) As of mid-1992, the program was approximately 50 percent complete, and the RCM results are taken to the floor and implemented as each systems analysis is completed and approved by the TMI staff.

2. *Westinghouse Electric Corporation, Sayreville combined-cycle plant.* Sayreville, located near Newark, New Jersey (and a sister plant, Bellingham, located near Boston, Massachusetts) is a 300 MW$_e$ combined-cycle plant. It uses two Westinghouse 501D5 combustion turbines exhausting into heat exchangers which supply steam to a single cylinder steam turbine as well as process steam to an off-site host. Each of the three turbines directly drives an electric generator. A unique feature in this plant is the use of an *air*-cooled condenser system to condense the steam exhausting from the steam turbine. The system uses 16 air-cooled, finned-tube heat exchange modules with air forced across the modules by 16 nine-blade, 30-ft-diameter fans. A simplified plant schematic is shown in Fig. 7.2. Our example uses the air-cooled condenser system which has the unique feature of being a relatively new system with little operating experience available for

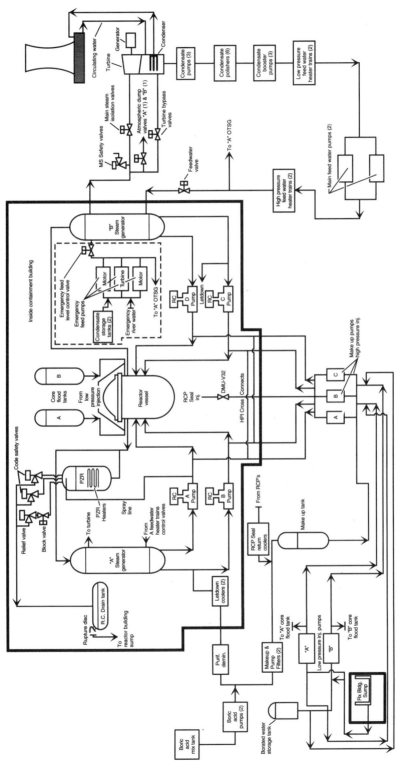

Figure 7.1 Simplified schematic—TMI-1 nuclear power plant. (*Courtesy of GPU Nuclear Corp.*)

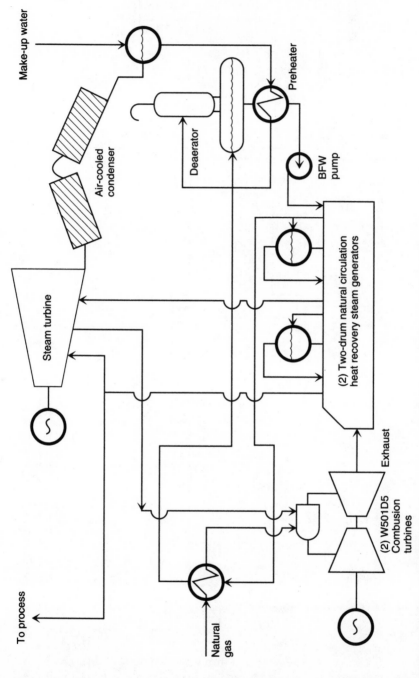

Figure 7.2 Simplified plant schematic—Sayreville combined-cycle plant. (*Courtesy of Westinghouse Electric Corp.*)

Make-up water

Air-cooled condenser

Preheater

Deaerator

BFW pump

Steam turbine

(2) Two-drum natural circulation heat recovery steam generators

Exhaust

(2) W501D5 Combusion turbines

To process

Natural gas

input to the analysis. Thus, OEM recommendations on PM tasks are the sole basis for comparison with the RCM findings.

3. *U.S. Air Force, Engine Test Facility (ETF).* The United States Air Force manages, and Sverdrup Technology operates, the largest and most extensive jet and rocket engine test facility in the world at the Arnold Engineering Development Center (AEDC) located 70 miles south of Nashville, Tennessee. Virtually every military and commercial jet engine in service today in the United States has been tested at AEDC during its development and qualification phases. The ETF is composed of three separate test plants (i.e., A, B, and C) for jet engines. The A and B plants have been in service since the late 1940s; the C plant has been in service since the mid-1980s and was designed with the capability to test very large high-bypass ratio jet engines, such as those for the 777 aircraft, at peak altitude and Mach number operating conditions. The Air Force has employed the RCM methodology as a basis for its PM program on tactical and strategic aircraft since the 1970s, and in a recent pilot project evaluated the use of RCM as an approach to improving maintenance efficiency and cost at the ETF. The refrigeration system for the A plant was one of two systems chosen for the pilot project, and is included here in our examples to illustrate an RCM application that has now gone beyond the power plant programs. It should be emphasized that this effort at AEDC is still being evaluated as to its cost-benefit potential for the ETF. The refrigeration system in question uses Freon as the cooling medium; gaseous Freon is compressed, cooled, and converted to liquid in a water condenser, passed through an evaporator for further cooling, and finally sent to a Freon-to-air heat exchanger where the test cell air can be cooled as low as $-70°F$. A simplified schematic is shown in Fig. 7.3.

4. *Florida Power and Light Company, Port Everglades fossil plant.* The Port Everglades Plant, located in Fort Lauderdale, Florida, is a four-unit fossil plant using both oil and gas fuel in all units. Two of the larger installations are 400-MW_e units, and in 1987 were the subject of the pilot RCM program at FP&L for their fossil generation facilities. (Recall that in 1983 the FP&L Turkey Point nuclear power plant was the first RCM pilot study for the utility industry.) The Combustion Air Delivery System (CADS) was chosen for the pilot fossil program. The CADS is the system which initially introduces air to the plant via large forced draft fans. The system then passes the air through the air preheater for delivery to the boiler windbox. The CADS schematic was previously shown in Chap. 5, Fig. 5.6, and a simplified plant schematic of a typical fossil unit is shown in Fig. 7.4. Since completion of the pilot study, selected findings have been implemented on the unit. Extensive RCM analyses were later conducted on 800 MW_e units, and implementation of these results is under review as this book goes to press.

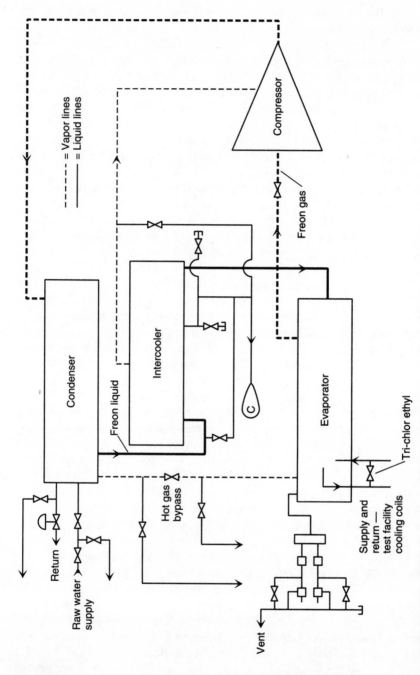

Figure 7.3 Simplified schematic—ETF refrigeration system. (*Courtesy of USAF/Sverdrup Technology, Inc.*)

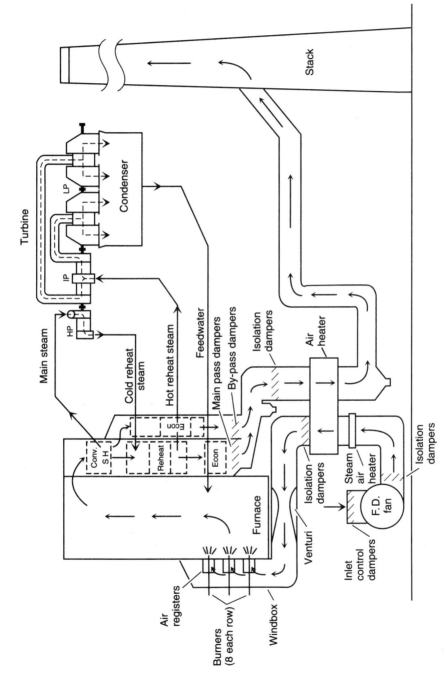

Figure 7.4 Typical fossil plant schematic. (*Courtesy of Florida Power & Light Co.*)

149

5. *Swimming pool.* This is the systems analysis example that was developed in Chap. 6. It is included in the overview here as a point of calibration and comparison for the other examples.

Typical systems analysis profiles. The RCM systems analysis profiles for the systems described previously are shown in Fig. 7.5. These profiles consist of summary counts of items that were developed at key points throughout the analysis process; the PM task profiles themselves are further discussed in the next subsection. There is really no "right" or "wrong" profile to expect. Each system tends to be unique in its own way; even with like systems each analyst may develop the analysis with some differences that ultimately lead to slightly different profiles. However, if the analysts have done a thorough and correct evaluation, one can generally expect to see the number of failure modes and PM tasks produced to be very similar for near-identical systems. Clearly, there is a wide range of different system complexities involved in the six examples in Fig. 7.5, ranging from simple with the swimming pool water treatment system to complex with the nuclear power plant main feedwater and instrument air systems.

You will note that all six systems, irrespective of complexity, were further divided into three, four, or five functional subsystems as was suggested in Chap. 5, Sec. 5.4. This simplified the analysis process considerably, even for the swimming pool evaluation. But, from here on, each system develops its own rather unique profile. Even then, however, we can observe some interesting similarities that tend to represent almost universal profile characteristics of the systems analysis process:

1. The ratio of functional failures to functions hovers around 2:1. In Fig. 7.5 the range is 1.4:1 to 2.6:1.

2. The ratio of failure modes to functional failures is usually in the 4:1 range, on average. Here, we see a low of 2.1:1 to a high of 11.7:1. Of course, the absolute number of failure modes for a given system can vary tremendously, and this is usually related to system complexity in terms of equipment count as well as the cross-functional role that the equipments play (i.e., the component failure modes may be involved in several different functional failures).

3. The ratio of failure modes to logic tree to failure modes is always less than 1:1. This ratio indicates the degree of redundancy available in the system as well as the amount of noncritical failure modes that are present in the system. (Recall from Chap. 5, Sec. 5.6 that we do *not* carry a failure mode forward to the LTA if its effect in the FMEA is "local only.") The water treatment system has the highest ratio at 0.89:1, and the refrigeration system has the lowest ratio at 0.30:1.

Source	Systems	No. of funct. subsystems	No. of functions	No. of funct. failures	No. of failure modes	No. of failure modes to logic tree	No. of RCM-based PM tasks
Nuclear power plant	Main feedwater[a]	3	69 (2.1)	145 (5.6)	806 (.74)	598 (.38)	230
Nuclear power plant	Instrument[a] air	5	136 (1.4)	187 (2.3)	433 (.88)	380 (.62)	235
Fossil power plant	Air-cooled[b] condenser	3	11 (2.6)	29 (11.7)	340 (.42)	142 (1.25)	178
Air conditioning plant	Refrigeration[c]	3	13 (2.3)	30 (3.7)	110 (.30)	33 (.97)	32
Fossil power plant	Combustion air delivery[d]	4	49 (1.7)	83 (2.1)	175 (.61)	106 (1.17)	124
Swimming pool	Water treatment	3	8 (2.1)	17 (2.2)	38 (.89)	34 (.91)	31

( ) = Ratio of right-hand column value to left-hand column value.
[a] GPU Nuclear–Three Mile Island-1
[b] Westinghouse–Sayreville (combined cycle)
[c] U.S. Air Force–engine test facility
[d] Florida Power & Light–Port Everglades (oil/gas)

Figure 7.5 RCM systems analysis profiles.

Thus, the former has little redundancy and/or noncritical failure modes, while the latter is just the opposite. As a rule of thumb, when the ratio is less than 0.5:1, there is a large amount of redundancy in the system to protect the functions.

4. Finally, as we review the number of RCM-based PM tasks that were specified for each system, we should not expect to see that number (including run-to-failure) equal the number of logic tree failure modes. Depending on how the counting is done, there could be three reasons why the task count is greater or less than the failure mode count: (1) a failure mode may have appeared in more than one functional failure, but only one PM task is eventually assigned to it; (2) several different failure modes may result in having the same PM task assigned to them (e.g., a single inspection or overhaul task may well cover multiple failure modes); and (3) during the sanity check of those failure modes that were dropped from the LTA, we may find that additional PM tasks are added to the list. Thus, it is virtually impossible to cite any general guideline as to how the failure modes will ultimately generate PM tasks. Each system is unique in that regard, as is amply demonstrated in Fig. 7.5, where the ratio of PM tasks to logic tree failure modes ranges from a low of 0.38:1 to a high of 1.25:1.

Typical PM task profiles.   We can extend our overview of the total PM task count that is shown in Fig. 7.5 (the far right-hand column) to include some additional statistical detail that further describes the PM task profiles. This is done in two ways.

First, we can take the total number of PM tasks, further divide them into task type, and then compare the current PM task structure with the RCM-based PM task structure as shown in Fig. 7.6. Again, there is no right or wrong profile, but some interesting observations can be made nonetheless. For example, notice that five of the six systems in our illustration had an increase in PM tasks resulting from the RCM analysis. This is not always the case, and in fact there is a gathering collection of experience to suggest that a decrease in total PM tasks is more likely the case. But even with an increase, we can often expect to see a decrease in PM costs simply because the more expensive TD overhaul tasks have been replaced with CD, FF, or RTF tasks, which have the net effect of greatly extending (even eliminating) the overhaul action. As cases in point, the TMI feedwater system is a net zero PM cost result and the instrument air system is a 900-hour PM labor-cost saving between reactor refueling outages—despite the overall increase in the number of PM tasks performed in both systems. Of course, even should the RCM process result in PM cost increases, our expectation is that corrective-maintenance cost reductions and accrual of cost-avoidance penalties (i.e., a reduction in forced outage rates) will more than offset such results. Also, note that the RCM tasks consistently take a greater advantage of CD and FF tasks, and with but one exception (refrigeration system) specifically call for RTF tasks as an integral part of the decision process. While one might want to think of the numbers in parentheses in Fig. 7.6 as equivalent to RTF, the author's experience leans heavily in the direction that *none* is more likely to mean that lack of PM tasks in the current program is really an oversight—not a deliberate decision. On that basis, clearly the current task structure contains a significant amount of oversight on all six systems.

Sometimes it is necessary to examine the task type statistics at the functional subsystem level in order to develop some sharper distinctions between the current and RCM task structures. Take, for example, the air-cooled condenser system which is further detailed in Fig. 7.7 with its three functional subsystems. In Fig. 7.6, it might appear that the 127 *none* directly translates to the 127 *RTF.* But Fig. 7.7 clearly shows that this is not the case. Also, note that at the functional subsystem breakout, two of the three subsystems had a net increase in PM tasks, while the third subsystem had an offsetting decrease. Thus, at the functional subsystem level, the differences between the OEM-recommended PM tasks and the RCM-based PM tasks become quite clear.

A second way to further describe the PM task structure is to construct the RCM task similarity profiles shown in Fig. 7.8. Rather than compare task type, as was done on Figs. 7.6 and 7.7, we now turn our attention to how the current and RCM-based PM tasks may have similarities or differences in their specific content (i.e., identical in content, similar in content, or completely different). This is done via the four classifications shown in Fig. 7.8. Again, there is no right or wrong profile, but some interesting observations can be made. If we group the first two classifications (the tasks are equal or quite similar), about 50 percent of the tasks are covered in four of the six systems (instrument air, refrigeration, combustion air delivery, and water treatment), thereby indicating that the RCM process agreed with about one-half of the existing PM program. Two systems, however, were exceptions to this at opposite ends of the scale. The main feedwater system had an 80 percent similarity value (i.e., the existing PM program was independently verified as being very good) while the air-cooled condenser system had only a 7 percent similarity value (i.e., RCM-based tasks were significantly different than the OEM recommendations). Neither result is too surprising; the TMI-1 PM program was revised and well structured during its six-year hiatus following the TMI-2 accident, and the air-cooled condenser system is quite new, with the RCM results demonstrating how useful the RCM process can be in addressing new systems and/or facilities. But the key areas of program optimization

	Main feedwater[a]		Instrument air[a]		Air-cooled condenser[b]		Refrigeration[c]		Combustion air delivery[d]		Water treatment	
	Current	RCM	Current	RCM	Current[e]	RCM	Current	RCM	Current	RCM	Current	RCM
Time-directed	162	188	103	169	18	14	1	11	43	39	5	9
Condition-directed	16	18	9	23	3	14	9	10	0	32	0	6
Failure-finding	18	19	21	33	30	23	6	11	24	29	0	3
Run-to-failure (none)	(34)	5	(102)	10	(127)	127	(16)	0	(57)	24	(26)	13
Total (RCM Δ)	230	230	235	235	178	178	32	32	124	124	31	31
		(+29)		(+92)		(0)		(+16)		(+33)		(+13)

[a] GPU Nuclear–Three Mile Island-1
[b] Westinghouse–Sayreville (combined cycle)
[c] U.S. Air Force–engine test facility
[d] Florida Power & Light–Port Everglades (oil/gas)
[e] OEM Recommendation

Figure 7.6 RCM task type profiles.

	Cooling subsystem		Storage subsystem		Air removal subsystem	
	OEM	RCM	OEM	RCM	OEM	RCM
Time-directed	10	5	6	6	2	3
Condition-directed	0	7	1	5	2	2
Failure-finding	2	10	0	5	28	8
Run-to-failure (none)	(21)	11	(47)	38	(59)	78
Total (RCM Δ)	33	33	54	54	91	91
		(+10)		(+9)		(−19)

Figure 7.7 RCM task type profile—air-cooled condenser system (Westinghouse–Sayreville, combined cycle).

derive mainly from the last two classifications in Fig. 7.8, where the RCM process either recommends PM tasks where nothing currently exists—or, conversely, recommends the outright deletion of current tasks as serving no particularly useful purpose. In the latter category, we see a demonstration of the fact that, on average, 5–10 percent of an existing (non-RCM) PM program is likely to fail the applicable and effective criteria which is the fourth key principle of RCM.

All of these profiles (Figs. 7.5 to 7.8) are based on statistics that reflect counts at the failure mode level. When these results are translated into specific PM actions that need to be taken on *components,* the picture changes somewhat, since we are now identifying several failure modes with a single component. Figure 7.9 illustrates this point for the air-cooled condenser system, where the original 340 failure modes (Fig. 7.5) and 178 RCM-based PM tasks (Fig. 7.6) come from 86 components. The interesting point on Fig. 7.9 is that 50 of the 86 components, or 58 percent, are in the RTF category. Since this system was designed with a large share of redundancy, it is not surprising to see this latter result.

7.2 Power Plant RCM Analyses

"Vertical slice" examples. In the previous section, we have seen some actual RCM analysis results that were presented in an overview fashion with the use of summary statistics. As a further illustration of RCM applications, we will now examine a *vertical slice* from three systems analyses in order to demonstrate the use of the RCM process described in Chap. 5 as it was employed on three different power plant systems. The vertical slice will trace the origins and resulting PM action(s) of a single failure mode through the rigors of Steps 3 to 7 of the systems analysis process. In so doing, we assume that Steps 1 and 2 were performed as described in Chap. 5 (and they were) so that we concentrate on those Steps that are germane to the four RCM princi-

	Main feedwater[a]	Instrument air[a]	Air-cooled condenser[b]	Refrigeration[c]	Combustion air delivery[d]	Water treatment
RCM task equals current task.	175	65	5	5	32	4
RCM task equals current task with some modification.	9	54	7	11	37	4
RCM task recommended, but no current task exists.	34	102	136[e]	16	42	7
Current task exists, but not recommended by RCM.	12[f]	16[f]	30	0	13	0
Total	230	237	178	32	124	15

[a] GPU Nuclear–Three Mile Island-1
[b] Westinghouse–Sayreville (combined cycle)
[c] U.S. Air Force–engine test facility
[d] Florida Power & Light–Port Everglades (oil/gas)
[e] Includes 9 tasks where RCM task is different from OEM task
[f] Includes current tasks to be continued but not supported by RCM

Figure 7.8 RCM task similarity profiles.

	Cooling subsystem		Storage subsystem		Air removal subsystem	
	RCM	OEM	RCM	OEM	RCM	OEM
Time-directed	1	4	4	3	3	0
Condition-directed	1	0	2	0	2	0
Failure-finding	5	1	5	0	8	22
Run-to-failure	5	—	16	—	29	—
None provided	—	10	—	24	—	16
Combined[a]	4	1	1	1	0	4
Total	16	16	28	28	42	42

NOTE: A component is assigned to indicate the presence of at least one active task—i.e., TD, CD, or FF. If *all* failure modes are RTF, only then is the component assigned to "RTF." Likewise for "none provided."

[a] If there are two different types of *active* tasks assigned to a component (e.g., TD and FF), then it is put into "Combined."

Figure 7.9 Component task profile—air-cooled condenser system (Westinghouse–Sayreville, combined cycle).

ples. Of course, we have already seen a complete RCM systems analysis in Chap. 6, so these illustrations are intended to additionally illustrate how the process works in more complex situations. The three power plant systems presented here are as follows:

1. Instrument air system: TMI-1 nuclear plant

2. Air-cooled condenser system: Sayreville combined-cycle plant

3. Turbine system: 800-MW$_e$ fossil plant (Martin plant at Florida Power and Light Co.)

The vertical slice for these systems is developed in Figs. 7.10, 7.11, and 7.12, respectively. In each case, only that portion of the analysis process that ultimately leads to the RCM-based PM task(s) for the selected failure mode is shown (therefore, the term *vertical slice*). Some comments on each vertical slice are given in the following paragraphs.

Instrument air system. This RCM analysis was performed concurrently with the introduction to service of some major system modifications to the normal instrument air functional subsystem. Thus, the RCM results were employed to specify the new PM tasks that were required. We see two such tasks that were derived from the failure mode in the vertical slice in Fig. 7.10 (a total of 90 new tasks resulted, overall, from the modifications). The example failure mode here, desiccant exhausted, was passed to the LTA since it produced both system and plant effects, but the LTA led to the relatively benign classification of D/C. Even then, the effectiveness of the candidate-applicable PM

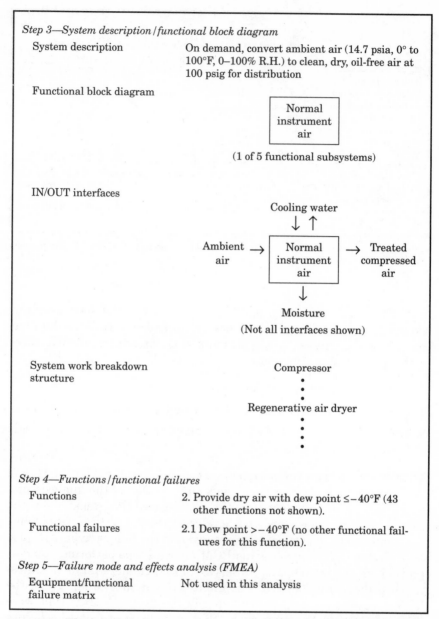

Step 3—System description / functional block diagram

System description   On demand, convert ambient air (14.7 psia, 0° to 100°F, 0–100% R.H.) to clean, dry, oil-free air at 100 psig for distribution

Functional block diagram

> Normal
> instrument
> air

(1 of 5 functional subsystems)

IN/OUT interfaces

Cooling water
↓ ↑

Ambient → | Normal instrument air | → Treated compressed air

↓

Moisture

(Not all interfaces shown)

System work breakdown structure

Compressor
•
•
•
Regenerative air dryer
•
•
•

Step 4—Functions / functional failures

Functions   2. Provide dry air with dew point $\leq -40°F$ (43 other functions not shown).

Functional failures   2.1 Dew point $> -40°F$ (no other functional failures for this function).

Step 5—Failure mode and effects analysis (FMEA)

Equipment/functional failure matrix   Not used in this analysis

Figure 7.10 "Vertical slice"—Instrument air system (GPU Nuclear–Three Mile Island-1).

FMEA

Equipment	Failure mode	Failure cause	Failure effects			LTA
			Local	System	Plant	
IA-Q-Z Air dryer	Desiccant exhausted	Normal wear	Wet desiccant	Wet air discharged into system. Long-term damage accrual.	Degraded operation due to water intrusion into equipment.	Y

Step 6—Logic tree analysis

Failure Mode = Desiccant exhausted

1. Evident: No
2. Safety: No
3. Outage: No

Category D/C
(minor economic problem)

Step 7—Task selection

Applicable task candidates

1. Monitor dew point continuously and alarm at −30°F (CD)
2. Inspect condition and color annually (TD)
3. RTF

Effectiveness factors

Both No. 1 and No. 2 are considered to be cost effective in order to avoid water damage to equipment. RTF not considered to be cost effective.

RCM decision

Perform both continuous dew point monitoring with alarm at −30°F, and the annual inspection. Plan to delete the annual inspection after operational verification of the dew point monitoring has been achieved.

Current task

New equipment. No previous PM tasks were in place.

Figure 7.10 "Vertical slice"—Instrument air system (*Continued*)

tasks was so superior to the RTF option that both PM tasks were incorporated in order to preclude the potential of severe water damage to various equipments. This example, incidentally, also demonstrates one of the reasons why we must continue periodically to review our baseline RCM tasks in a living RCM program (see Chap. 8, Sec. 8.4).

Air-cooled condenser system. As noted previously, this system is a relatively new design for use in power plants (water is usually the cooling medium). Thus, most of the OEM recommendations for PM tasks were derived from component applications that were quite different from those experienced here. For example, the fans in this system are

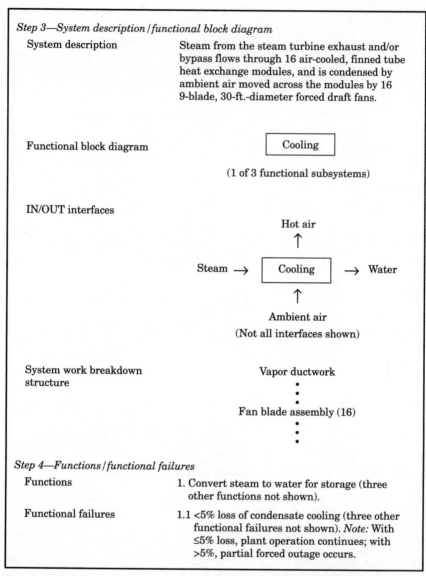

Step 3—System description / functional block diagram

System description   Steam from the steam turbine exhaust and/or
                       bypass flows through 16 air-cooled, finned tube
                       heat exchange modules, and is condensed by
                       ambient air moved across the modules by 16
                       9-blade, 30-ft.-diameter forced draft fans.

Functional block diagram

Cooling

(1 of 3 functional subsystems)

IN/OUT interfaces

Hot air
↑

Steam →   Cooling   → Water

↑
Ambient air
(Not all interfaces shown)

System work breakdown
structure

Vapor ductwork
•
•
•
Fan blade assembly (16)
•
•
•

Step 4—Functions / functional failures

Functions   1. Convert steam to water for storage (three
                       other functions not shown).

Functional failures   1.1 <5% loss of condensate cooling (three other
                       functional failures not shown). *Note:* With
                       ≤5% loss, plant operation continues; with
                       >5%, partial forced outage occurs.

Figure 7.11 "Vertical slice"—air-cooled condenser system (Westinghouse–Sayreville, combined cycle).

suspended some 70 feet aboveground, and are accessible only from above on adjacent catwalks that even then make any work on the fan blades very difficult and perhaps even dangerous. Thus, PM tasks that are focused on the avoidance of the example failure mode, blade separation, can be extremely cumbersome, time-consuming, and expensive

Step 5—Failure mode and effects analysis (FMEA)

Equipment/functional
failure matrix

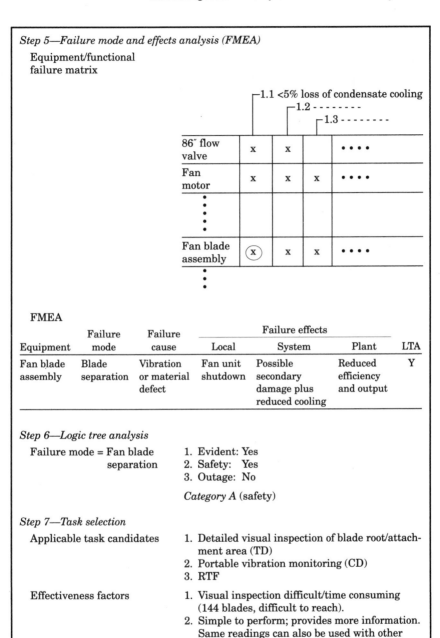

┌ 1.1 <5% loss of condensate cooling
  ┌ 1.2 - - - - - - - -
    ┌ 1.3 - - - - - - - -

	1.1	1.2	1.3	
86″ flow valve	x	x		• • • •
Fan motor	x	x	x	• • • •
⋮				
Fan blade assembly	(x)	x	x	• • • •
⋮				

FMEA

			Failure effects			
Equipment	Failure mode	Failure cause	Local	System	Plant	LTA
Fan blade assembly	Blade separation	Vibration or material defect	Fan unit shutdown	Possible secondary damage plus reduced cooling	Reduced efficiency and output	Y

Step 6—Logic tree analysis

Failure mode = Fan blade separation

1. Evident: Yes
2. Safety: Yes
3. Outage: No

Category A (safety)

Step 7—Task selection

Applicable task candidates

1. Detailed visual inspection of blade root/attachment area (TD)
2. Portable vibration monitoring (CD)
3. RTF

Effectiveness factors

1. Visual inspection difficult/time consuming (144 blades, difficult to reach).
2. Simple to perform; provides more information. Same readings can also be used with other components (e.g., motor, gearbox).
3. RTF *not* an option when safety is involved.

Figure 7.11   "Vertical slice"—air-cooled condenser system (*Continued*)

| RCM decision | Use *portable vibration monitoring*. Also, perform initially at *3-month intervals*, but initiate age exploration to increase periodicity after baseline signature trends are established. |
| OEM recommendation | Check all bolts and nuts on all 144 blade-hub joints at 6-month intervals (TD). *Note:* Similar to candidate no. 1 above which was rejected in favor of portable vibration monitoring. |

Figure 7.11 "Vertical slice"—air-cooled condenser system (*Continued*)

if any work is required directly on the fan blade–hub attachment area. However, the failure mode is a safety-critical classification of A, and some action is mandatory. The final RCM decision to employ a rather simple vibration-monitoring task was clearly the preferred option in this system.

Turbine system. This example is a classic situation where the current PM task had been in place for years, so "it must be the right thing to do." The failure mode in question, a throttle valve that fails to open, results initially in the relatively benign classification of D/C, since redundancy can negate any serious effect of a single valve failure. However, here is a case where historical evidence was brought to the table to highlight the fact that a second valve failure (on the heels of a hidden first failure) had occurred with sufficient frequency to move the classification to D/B. In the ensuing task selection process, we see both the emergence of a new failure-finding task to eliminate the hidden (surprise) element as well as a greatly expanded time-directed task to inspect critical areas of the valve where failure had historically been observed.

Task selection examples. As a final illustration of power plant RCM analysis results, we will review some selected task comparisons that were developed from the RCM systems analysis process conducted on eight different systems. The specific task comparisons in our examples are shown on Fig. 7.13.

These eight examples help us to further understand the power of the RCM process by demonstrating some important facets of PM optimization that were achieved:

1. Some fair percentage of current PM tasks are simply not applicable tasks. Item 1 illustrates this point. In the air-cooled condenser system, the makeup water for the system is carefully controlled to avoid metal corrosion, scaling, chemical attack, etc. Thus, an expensive cleaning overhaul of the vacuum deaerator condenser would simply be a waste of PM resources!

2. While some PM tasks may be applicable tasks, they clearly are not always effective tasks. Item 2 illustrates this point. The current

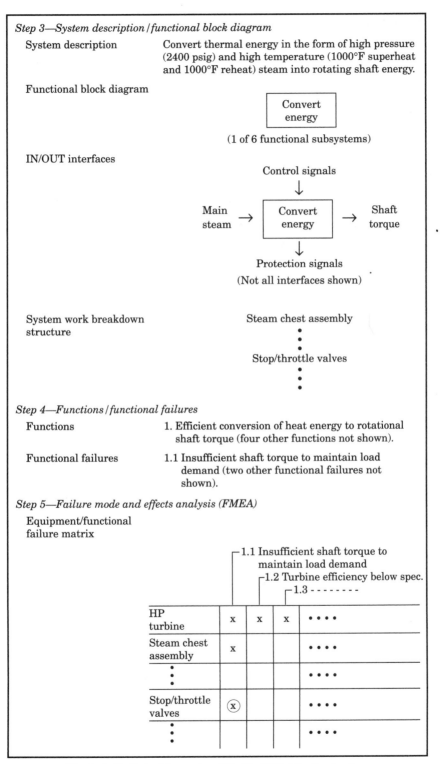

Step 3—System description / functional block diagram

System description — Convert thermal energy in the form of high pressure (2400 psig) and high temperature (1000°F superheat and 1000°F reheat) steam into rotating shaft energy.

Functional block diagram

Convert
energy

(1 of 6 functional subsystems)

IN/OUT interfaces

Control signals
↓

Main steam → Convert energy → Shaft torque

↓
Protection signals
(Not all interfaces shown)

System work breakdown structure

Steam chest assembly
•
•
•
Stop/throttle valves
•
•
•

Step 4—Functions / functional failures

Functions — 1. Efficient conversion of heat energy to rotational shaft torque (four other functions not shown).

Functional failures — 1.1 Insufficient shaft torque to maintain load demand (two other functional failures not shown).

Step 5—Failure mode and effects analysis (FMEA)

Equipment/functional failure matrix

	1.1 Insufficient shaft torque to maintain load demand	1.2 Turbine efficiency below spec.	1.3 - - - - - - - -	
HP turbine	x	x	x	• • • •
Steam chest assembly	x			• • • •
• • •				• • • •
Stop/throttle valves	(x)			• • • •
• • •				• • • •

Figure 7.12 "Vertical slice"—turbine system (Florida P&L–Martin plant).

FMEA

Equipment	Failure mode	Failure cause	Failure effects			LTA
			Local	System	Plant	
Stop/throttle valve	Fails to open	Broken stem due to cyclic stress	Steam flow blockage	Steam flow mismatch (short duration)	None— other valves adjust to accommodate	Y

Step 6—Logic tree analysis

Failure mode = Fails to
                open

1. Evident: No
2. Safety: No
3. Outage: No (see note)

Category D/C (minor economic problem)

Note: Since multiple broken stems have been experienced, this LTA was upgraded.

Category D/B (can eventually lead to plant outage)

Step 7—Task selection

Applicable task candidates	1. Monitor steam chest metal ΔT on daily basis (FF) 2. Inspect stems for evidence of cracking or deterioration at turbine overhaul every 6 years (TD) 3. RTF
Effectiveness factors	1. Easy to implement, thus very cost-effective. 2. Can be easily done in conjunction with other necessary work. 3. Not cost-effective.
RCM decision	Perform both the ΔT measurement and the detailed inspection at the opportunity presented.
Current task	No FF task in place. Current 6-year inspection very superficial (e.g., does not include y–x seat shoulder areas and tread to shoulder transition area with dye penetrant inspection which RCM task requires).

Figure 7.12 "Vertical slice"—turbine system (*Continued*)

practice was to replace the transformer fan bearing every three years. The RCM analysis, however, revealed three key points not previously considered: (1) the fan bearings, upon examination at removal, were seldom in need of replacement, (2) the fans are redundant, thus the cooling function was not lost even if one fan failed, and (3) a spare fan was readily available within two hours. Again, unnecessary and wasted resources! RTF was the solution.

System title	Component/ failure mode	RCM program RCM-based task	Freq.	Existing program Current task	Freq.	Comment
1. Air-cooled condenser	Vacuum deaerator condenser/internal fouling	RTF	—	Clean shell and tube sides (TD)	1 year	System water quality makes fouling implausible.
2. Electrical	Transformer fan/ seized bearing	RTF	—	Replace bearing (TD)	3 years	Fans are redundant. Spares are available.
3. All systems	Manual valves/leaks	Visual inspection during walkdown (FF)	Daily	Repacking (TD)	1 year	Costly to repack, often causes leaks due to repacking errors.
4. Main generator	Generator/journal bearing fails	Perform oil analysis (CD)	1 month	None	—	Monitoring very cost effective.
5. Circulating water	Pump/seized coupling	Inspect for wear and lubricate (TD)	2000 hours	Repack coupling (TD)	1 year	History of wear and need for more frequent lubrication.
6. Feedwater	Pump/suction strainer plugged	Record pressure drop across strainer (CD)	1 month	Clean strainer (TD)	1 month	Infrequent clogging history. Operations can easily monitor pressure drop.
7. Condensate	Heater drain motor/seized bearing	Monitor bearing temperature and vibration (CD) Perform oil analysis (CD)	3 months 18 months	Overhaul (TD)	6 years	Overhaul history void of problems. Monitoring very cost effective.
8. Electro-hydraulic control	Load unbalance amplifier/out of adjustment high	Perform calibration (TD)	18 months	Perform calibration (TD) (but does not include load current card)	18 months	Minor modification to current task was very cost effective.

Figure 7.13 Task selection—examples.

3. A variation on this same theme is illustrated in item 3. The common perception is that *all* fluid system valves need periodic (usually annual) repacking. However, most *manual* valves do not support highly critical functions, nor is a leak in such valves a serious concern. So why repack them if they are not leaking?? Worse yet, repacking will frequently induce a leak where such was not present had well-enough been let alone. (How often have you had an oil/filter change on your automobile engine only to discover, when you have driven back home, that there is an oil leak on your driveway? The moral is that PM tasks frequently involve risks, as discussed in Chap. 2, Sec. 2.5.) A failure-finding task was the more appropriate solution.

4. The current PM program often neglects several failure modes which can be addressed in a very cost-effective fashion. Item 4 illustrates this point by incorporating an oil sampling and analysis task to detect incipient failure conditions in the journal bearing. Although such failures in the main generator are infrequent, their occurrence is a surprise in the absence of this CD task, and will usually result in an extended forced outage. With proper warning, the bearing replacement can usually be done during a regular scheduled outage—or, in the worst case scenario, can be anticipated and planned for properly to keep the forced outage to a fraction of the time that would otherwise accrue in a surprise situation.

5. Even if the current PM program recognizes important failure modes, it will not always specify the correct periodicity for an applicable task. Such is the case in Item 5 where the annual repacking was a good example of the too-little-too-late adage. A review of the corrective maintenance history made it clear that insufficient lubrication was causing premature wear. The simple introduction of a more frequent lubrication task, based on the age-reliability indications in the corrective maintenance actions, prevented the seizure failure mode.

6. One of the major problems that is repeatedly observed in current PM programs is the excessive use of intrusive TD overhaul tasks. The age-reliability relationships are usually an unknown, yet the overhaul tasks are continued without a conscious review of what (if anything) is being derived from the expenses incurred. Items 6 and 7 are illustrative of this point. In both cases, intrusive TD tasks were being performed while the historical records reflected little, if any, benefit from the actions. Thus, CD tasks were incorporated, at a fraction of the prior cost, to monitor for the incipient failure condition.

7. Finally, we have the situation where the current PM task is correct but, with a slight modification, can be adjusted or extended to increase its applicability and/or effectiveness. Item 8 illustrates this

point. The amplifier in question was receiving the appropriate calibration, but inadvertently had not included the load current card in the calibration task. So the existing TD task was very simply modified to extend the calibration to this card.

To summarize, the preceding examples illustrate six of the most common areas where the RCM process has been instrumental in optimizing PM programs:

1. Deleting PM tasks that are not applicable (Fig. 7.13, item 1)

2. Deleting or redefining PM tasks that are not effective (Fig. 7.13, items 2 and 3)

3. Adding PM tasks to address neglected, but important, failure modes (Fig. 7.13, item 4)

4. Revising applicable PM tasks to reflect a correct periodicity (Fig. 7.13, item 5)

5. Replacing expensive, nonapplicable overhaul tasks with cost-effective CD tasks (Fig. 7.13, items 6 and 7)

6. Modifying existing applicable PM tasks to increase their effectiveness (Fig. 7.13, item 8)

7.3   Peripheral Benefits Realized from RCM Process

As our experience with the RCM process developed, we discovered a very pleasant surprise that was happening on virtually every systems analysis that was conducted. This surprise was the accrual of a series of peripheral technical and cost benefits that simply fell out from the rigors and thoroughness that is inherent to the RCM process. These peripheral "gratis" benefits were occasionally so significant that they alone were estimated to have paid for the cost of the systems analysis several times over. When people speak of the cost-benefits derived from an RCM program, these peripheral benefits are almost never included in their figures since future predictability of such benefits is elusive, and thus difficult to claim as expected credit. Nonetheless, it is the author's view that such peripheral benefits will continue to occur, and thus represent just one more convincing argument as to why management should embrace the RCM methodology.

To drive this point home, several of the actual benefits that have been experienced are briefly discussed in order to illustrate the type of gratis fallout that you might see in your RCM program. We can put these benefits into one of five categories as a means of further identifying their favorable impact:

- Operational impact (OI)
- Safety impact (SI)
- Logistics impact (LI)
- Configuration impact (CI)
- Administration impact (AI)

All of them have a favorable cost impact, but this is not specified here (with one exception) because these items were not cost-quantified, as a rule, by the organizations involved. In retrospect, this was probably a mistake that should be corrected in future programs.

The list of examples is presented below, and their area of impact is indicated parenthetically in each case.

1. When completing the SWBS in Step 3 of the systems analysis process, the analyst discovered that the latest revision to the system drawings contained a change in the model number for one of the system components. This led to an investigation to verify which model number should be used in the analysis. The final answer confirmed that the new model number was correct—but it also revealed that several spares for the *old* model number had inadvertently been retained in the stores inventory. This old inventory had been mistakenly held for some four years at an estimated cost of $75,000 to cover taxes, storage, and administrative expenses, before the RCM program triggered the actions necessary to dispose of the useless spare parts. (LI and AI.)

2. When the analyst was preparing the comparison of the RCM versus current PM tasks in Step 7, it was necessary to research the documentation defining the current PM actions. In so doing, it was discovered that the operations and maintenance departments were both performing some identical PM actions on the same component—but with different periodicities such that neither group ever discovered the duplication that was regularly occurring. The RCM process had resulted in a recommendation to do essentially the same task, so its implementation was instrumental in assigning the task responsibility solely to the maintenance department. (OI and AI.)

3. There have been several instances where the analysts have discovered that a system contains one or more components that serve no useful role whatsoever in supporting the functions that have been identified for the system. Worse yet, this group of components frequently contains failure modes which can result in plant forced outages should they occur. The analysts tend to discover such situations at two points in the process: first, during the compilation of the SWBS in Step 3

where a detailed correlation is developed between the functional sub-systems and the system P&ID and, second, during the development of the equipment-functional failure matrix in Step 5, where it is first discovered that an empty set exists in the matrix. As a rule, these components are removed from the system at the earliest possible convenience, which is usually the first scheduled outage after support engineering has verified the analyst's finding. In many cases, the components involved in these problems are valves of one type or another (flow control, check, block, etc.) with solenoid valves, limit switches, instrumentation, and other peripheral items also involved. (OI, SI, LI, CI, AI.)

4. The rigors of the systems analysis process at virtually every step along the way require a detailed review of the system documentation. This, in turn, very often identifies either missing or erroneous information in the system baseline definition records. Such information includes corrections or additions to system P&IDs, system equipment and parts lists, configuration control files, maintenance management information system files, equipment tags, etc. Correction of such information is particularly significant in nuclear power plants or any facility where safety systems are key players in the everyday operations. (OI, SI, CI.)

5. The rigors of the process are also capable of identifying simple design enhancements that can eliminate failure modes, failure effects, and/or PM tasks. Consider the following examples: (1) upgrading an analog to a digital control device or instrument, (2) adding manual isolation valves to allow repair/replacement of failed items in redundant configurations (the author never ceases to be amazed at the frequency of such simple omissions in the original design), and (3) installation of the capability for manual addition of chemicals to automatic water-conditioning systems that fail. (OI, SI, CI.)

6. If the plant or system is *new,* the RCM process offers some excellent opportunities to correct deficiencies that were overlooked during the preoperational punch-list walkdown and checkout. (And yes, these deficiencies do occur in the real world in spite of TQM, zero defects, TLC, and all of the other popular buzzwords that currently dominate our corporate management philosophy!) Two examples illustrate this point: (1) a system walkdown during Step 2 of the systems analysis process discovered that valves needed to activate an air-removal subsystem had not been opened (the immediate effect was inconsequential, but the long-term effect was potentially very damaging due to corrosion concerns) and (2) improper piping connections had made a water chemistry analyzer inoperative. Notification to operations corrected both situations in a timely fashion. (OI, CI.)

7. One RCM program used Step 7—task selection—as an opportunity not only to specify the RCM-based PM task recommendations, but

also to further highlight where special instructions, cautions, and/or training requirements were needed to successfully implement the tasks. This addition to the task selection process is particularly noteworthy in light of the industrywide deficiencies associated with the lack of good documentation on standard maintenance practices and procedures. (SI, AI.)

8. Occasionally, Step 5—FMEA—will discover a heretofore unrecognized failure scenario that could initiate or directly produce serious operational or safety consequences. The likelihood of such a discovery during the RCM process is admittedly small, but the fact is that the FMEA presents an excellent opportunity to revisit a variety of operational conditions that in many older plants have not been reviewed for years. There have been recorded incidents of such discoveries in the systems analysis process and, in one case, the discovery was made even after the completion of an extensive Probabilistic Risk Assessment (PRA) on the system in question. (OI, SI.)

In addition to the preceding specific examples, virtually every organization that has conducted an RCM program has recognized the value of the systems analysis process as a training ground for system engineers. The training, in fact, is so comprehensive that the analysts often become the recognized "resident system expert" for those systems which have experienced the RCM process. Another related recognized value is the use of the systems analysis information in operator training exercises wherein, for example, failure scenarios developed in the FMEA would be used as simulator inputs to test operator response to plant transient or upset conditions. Both of these training benefits need to be more thoroughly developed by management in order to realize the full potential that they offer.

8

Implementation— Carrying RCM to the Floor

Upon completion of the Seven-Step systems analysis process described in Chap. 5, we must now take the final and crucial action to realize the fruits of our efforts. We must carry the RCM recommended tasks to the floor. That is, we must accomplish the Task Packaging that was first introduced in Chap. 2, Fig. 2.1.

The Task Packaging effort has frequently proven to be a difficult task to successfully and efficiently accomplish. "Ten key lessons to remember" (see Sec. 8.1.1) have been developed to help the maintenance practitioner to deal with the issues and problems of Task Packaging in a timely manner. Two of the lessons, "Organizational considerations" and "Using CD and FF tasks," are discussed in some detail to highlight critical considerations such as approval and funding issues, buy-in, resource allocation, and the role of operations and technical support in the PM program.

The role of the computer in the RCM process is outlined, indicating its usefulness in automating input data and establishing a database of RCM information. However, it is emphasized that RCM is an engineering process that requires human knowledge, experience, and judgment. Thus the computer is an aid to data and information collection, storage, and retrieval, but it does not play any direct role, per se, in the systems analysis process.

The question of task periodicity is addressed with emphasis on the need to develop age-reliability statistics to accurately define task intervals. Since such statistics are often unavailable, the process known as *Age Exploration* is described as an empirical approach to refining the judgments that are used to initially establish task intervals.

Finally, the need to maintain a Living RCM Program is presented. Factors such as errors in the original analysis, new/modified equipments or procedures, new predictive maintenance technology, and the accumulation of cost-benefit data are sighted as reasons for the periodic review and update of the baseline RCM program.

8.1 Analysis to Implementation— Key Transition Factors

When we have completed Step 7 of the systems analysis process, we have specified the PM tasks which will optimize our PM program— that is, the tasks that will give us the best return for the resources invested. The rigors of the RCM process have defined the task content (i.e., What task?), and we have also made an estimate of the task frequency (When done?) such that the ideal PM program as originally discussed in Chap. 2, Fig. 2.1 has been specified. We must now complete the PM *Task Packaging* effort in Fig. 2.1 in order to take our optimized PM program to the floor. Several direct and indirect factors that are important to a successful Task Packaging effort are presented in this section.

Before moving on to that discussion, it is somewhat instructive to first point out that most of the utilities involved in an RCM program have to date encountered some level of difficulty in achieving Task Packaging. That is to say that they have found it difficult, for one reason or another, to take the RCM-based tasks to the floor. In some cases, these difficulties have been so severe that the RCM program has never been fully implemented. This situation was, quite honestly, a complete surprise to the author since it was assumed that the most difficult hurdle would be completing the systems analysis process. At first, these difficulties were also a complete mystery because implementing PM tasks (irrespective of their origin) seemed to be a straightforward activity that was already part of a power plant organization and infrastructure. As it turned out, this was not necessarily the case. Several reasons for the difficulty eventually emerged. Here are some of the observed impediments:

- The infrastructure used in processing a large number of PM changes had not been exercised for years, and nobody remembered how to do it. And starting over from ground zero seemed to be too time-consuming (especially in the reactive environments that all too frequently existed).

- Plant staff "buy-in" to the RCM program had not been achieved.

- The disconnect between operations and maintenance doomed the CD and FF tasks and, along with them, the entire program.

- Although the path to implementation was clear, no one had the time to follow it.

- Some CD tasks required new equipment and operator training, but resources for this were not available. So most, if not all, of the program was abandoned.

In most cases (but not always), the difficulties and impediments were overcome, and successful implementation was accomplished. But for those embarking upon their first RCM program, the caveat is quite clear—you must decide on how you will achieve Task Packaging from the outset, and plan accordingly, to avoid the preceding pitfalls, so that all of your effort and expense in the systems analysis process does not become just another dust collector on the shelf.

8.1.1   Ten key lessons to remember

Successful implementation means that we recognize the type of impediments that can occur, as well as the technical and programmatic issues that will arise, and deal with them in a timely fashion. This includes making them visible to all parties with a vested interest in such issues, and assuring that a stumbling block on just one or two relatively minor points will not doom the entire program.

While a sizable compendium of experience on Task Packaging has been collected, we have sorted this out here into the "ten key lessons to remember." This list, in the author's judgment, highlights the issues that most often can spell the difference between RCM program success and failure, depending on how they are handled.

Lesson 1: Equipment to function mind set.   We have touched on this subject before (e.g., in Chap. 5, Sec. 5.1), but it deserves reemphasis here. Failure to achieve some acceptance of the basic "preserve function" principle means that the maintenance staff never really understood why RCM was the better way to do things. Failing that, they will continue to hold to the premise that the old way of doing things (i.e., preserve equipment) is best, and they will find the means to assure that the old way prevails.

Lesson 2: Organization factors.   This issue is very important, and Sec. 8.1.2 is devoted to it.

Lesson 3: Buy-in.   Buy-in is the process whereby an individual or a group, responsible for carrying out some action, has been a party to the development and planning for that action. When buy-in is successfully accomplished, the people involved have made a direct contribution to the action plan and, implicitly, have accepted the plan as well as assumed some level of ownership in the plan. Without the essential ingredients of *acceptance and ownership,* it is highly improbable that a plant staff will feel motivated and compelled to implement anything— and that especially includes the recommended PM tasks from an RCM program. Achieving an appropriate level of buy-in to an RCM program is dependent upon several factors that deal with how the plant staff is integrated into the systems analysis process and the expected benefits therefrom. First of all, there must be a clear and visible endorsement for the RCM program from top management—usually at the vice-presidential level of the appropriate organization to which the plant or facility reports. However, do not be lulled into believing that the sales and education job stops there; it does not! If you succeed in obtaining top management endorsement, your job has really just begun because you now must do the same job, only better, with the plant staff. This includes the systems analysts who may or may not be members of the plant staff (see Sec. 8.1.2). This latter process is not a one-stop job, and chances are excellent that the sales and education process will continue over a long period of time—say, two years or more—to capture everyone in the plant organization who is germane to a completely successful RCM program. The approach to buy-in from the plant staff is multifaceted, and requires not only training seminars and one-on-one tutorials but, more importantly, it requires *involvement* in the systems analysis process as was noted throughout Chap. 5. This level of plant staff involvement not only injects valuable technical experience into the process, but it ultimately becomes the primary mechanism whereby ownership is instilled. The organizational approach to the RCM program also constitutes a vital factor in how the buy-in scenario plays out, and this is further discussed in Sec. 8.1.2.

Lesson 4: PM task procedures. In every RCM program, there are bound to be new and modified PM tasks that will require the generation of new or modified procedures before the task can be performed on the floor. The degree of formality and detail required of these procedures will vary considerably from plant to plant, with safety systems/equipment in a nuclear power plant probably representing the extreme requirement for such formality and detail. The problem encountered here is, quite simply, to determine who is responsible for preparing these procedures. Or, if responsibility can be clearly defined (which is rarely the case), is their time available to write the procedures (in between all of the "firefighting" that occupies such a large segment of the available work hours)? Thus, an RCM program plan and schedule must address, from the outset, just how this will be accomplished. You don't wait until the last minute, and just dump it on some unsuspecting group or individual. That approach does not work! Ideally, the plant organization has an identified group whose job is procedure upkeep, and the procedure issue will not become one of those unfortunate stumbling blocks. If this is not the case, then the systems analysts (with appropriate help—externally supplied, if necessary) are probably the next best course of action. The worst thing you could do is to dump the job on the maintenance supervisors and/or leads, who are probably not inclined to do such work, or if they are, will not have the time to do it.

Lesson 5: The run-to-failure shock.   The problem here has its origin in lesson 1, but is goes beyond the mind-set difficulty described in lesson 1. Even when "preserve function" is accepted, if the plant staff is not fully aware of the workings of the systems analysis process, the grease-under-the-fingernails maintenance person will still have real difficulty accepting the RTF decisions. Thus, it becomes additionally necessary to assure that the plant staff understands and appreciates the concept of "applicable and effective" and the vital role that this concept plays in deciding where to allocate the limited maintenance resources.

Lesson 6: Condition-directed and failure-finding tasks.   This fairly special topic receives an expanded discussion in Sec. 8.1.3.

Lesson 7: Challenging sacred cows.   Existing PM programs, which have usually evolved via an ad hoc process over the years, are prone to be saddled with a variety of tasks that originated external to the plant. These PM tasks come mainly from the OEMs, the insurance companies, and the regulators. Most of these tasks have taken on the aura of sacred cows which, by our definition here, means that they cannot be changed or deleted without the approval of some external (and all-knowing!) authority. Clearly, this issue has the makings of an impediment that may not even let you reach first base with an RCM program. There is really only one way to deal with this problem—and that is via an early and direct frontal attack. As a part of your initial sales and education effort, you must convince people that improvement and optimization will happen only when there is a level playing field where *nothing* is initially sacred. That is, let the chips fall where they may. Only then will the process give you the complete picture on how your PM program can be enhanced. Undoubtedly, the recommended RCM tasks will conflict with some percentage of the current PM tasks. But now you know where those differences reside and how serious they might be; and you are in the position of further deciding what, if anything, may be necessary in the way of additional interface to obtain external approval to make the change. Doing the latter could be an arduous process—especially with regulators. But if the effectiveness trade-off is warranted, then a decision to push for the RCM-based task is the right course of action to pursue. If you let sacred cows inhibit the RCM program from the outset (i.e., certain equipments or functions are off-limits), you will never know what could have been. Hopefully, you will encounter creative and inquisitive people who will not allow sacred cows to stand in the way.

Lesson 8: MMIS integration.   If the plant or facility already has a Maintenance Management Information System (MMIS), then the entire question of MMIS requirements in the RCM program becomes a relatively minor or nonexistent issue. But if the converse is the case, which is more likely with older plants (except for nuclear power plants), people could raise this as a stumbling block even though it may not be totally true. There is no question that an MMIS facilitates the efficient conduct of a PM program—any PM program. But with an RCM program, the efficient monitoring of CD task parameters (including an ability for automatic alerts), the compilation of PM and CM cost data for benefit analysis, and the tracking of Age Exploration programs, component histories for statistical analyses, and other related program measurements could become cumbersome in the absence of an MMIS. Thus, a decision to proceed with an RCM program could also involve the decision to acquire an MMIS if such is not already in place. In our modern age of reasonably priced PCs and the ready availability of proven software, this should not prove to be a difficult decision to make. It certainly should never become a reason for not embarking upon an RCM program.

Lesson 9: Labor and material adjustments.   Chapter 7 illustrated that an RCM program typically can be expected to introduce new (and often more technically sophisticated) CD tasks, and also to modify existing tasks, which frequently is the extension of overhaul intervals. Both of these changes will introduce labor and material adjustments to the current PM program. Labor adjustments will come primarily in the form of new or revised skill requirements for the craft personnel and technicians to carry out the CD tasks. Material adjustments may come in the form of new equipment and tool requirements, as well as a decreased spare parts inventory, as the effect of fewer component failures accrues over time. Here again, the effectiveness aspect of the decision process should help to guide decisions on the commitment of capital expenditures where such are warranted, and the positive effect of reduced inventories should be tracked and measured to reflect the resulting cost savings. The cost-benefit decisions should be quantitative in nature, and not relegated to the gut-feel and "I think" inputs which all too frequently are the modus operandi within today's business climate. Thus, the need to introduce new tools, equipment, and skills should never become an impediment when the return on such investments is clearly beneficial to the bottom line.

Lesson 10: Reduction in the plant staff. Discussions about labor adjustments deal not only with the issue of skill and training needs, but also with the question of workforce reductions at the plant. The reasoning goes along the line that if the RCM program is, in fact, capable of reducing PM costs on the systems where RCM is applied, then this must equate to a loss of jobs. To date, the evidence overwhelmingly indicates that such reductions do *not* occur. There are several reasons for this, but two factors seem to dominate. First, a major fraction of any PM savings that may be realized develops from either the extension or elimination of complex overhaul tasks—overhauls which are usually performed by vendors or at outside maintenance shops where plant staffing is not directly affected. Second, plant staffing is usually on the "mean and lean" side to start with; thus, the RCM program is aimed at obtaining the best possible productivity and plant availability for the costs incurred—not the reduction of the staff required to achieve this objective. The main concern here, if any, should be the possibility that new skills may be needed and the current staff may need to learn these skills or be bypassed as progress marches on. The current rage in corporate America is "right sizing" (a euphemism for staff reduction), but these actions are aimed at trimming fat, not muscle. The RCM program deals exclusively with muscle!

8.1.2 Organizational considerations

There is an old adage that it is the dispositions, personalities, and motivations of the people, not the structure of the organization in which they work, which ultimately determine project or product success or failure. Experience bears this out, in the author's view. But, by the same token, this experience also says that the particular version of organizational structure that is employed can be a significant factor in making success easy or difficult to achieve. For example, organizational structures usually determine lines of communication which can be short and simple or lengthy and complex; they also establish boundaries on areas of responsibility which can be either very broad, highly partitioned and restrictive, or even deliberately overlapping and competitive, where the "best ideas" emerge victorious. (I have worked in the latter organizational philosophy on occasion, and have frankly found it to be quite counterproductive to the ultimate product success, and sometimes even destructive of highly competent people who were inadvertently caught in its web.) Our objective here, however, is not the development of a lengthy discourse on organization concepts and practices but, rather, to briefly look at how typical organization structures and use of these structures might help or hinder successful RCM program implementation.

To keep it simple, we will look at two tiers of a utility-type structure, namely the corporate level and plant level, that reasonably represent the vast majority of organizations encountered by the author in his work with RCM programs. These structures also reasonably portray, at least in concept, the organizations that any product-oriented company might employ.

In the corporate structure shown in Fig. 8.1, our interests deal with the vice president of engineering support and vice president of generation since both of these positions may have roles and responsibilities that deal with one or more aspects of RCM. Frequently, the director of technical support, reporting to the vice president of generation, will also be a major player at the corporate-level involvement in RCM, and together with the director of power production, may actually represent their boss in any RCM decision-making situation. The point here is that corporate decisions on an RCM program usually involve at least two senior-level managers, and may involve as many as four of the top-level corporate managers. Thus, when seeking positive decisions on an RCM program at corporate headquarters, you need to understand who the essential decision makers will be; failure to include all of the right people in the selling process could automatically lead to an outcome that will doom your efforts before you even reach first base. In the plant-level structure shown in Fig. 8.2, our interests deal with the plant general manager as well as the superintendents of operations, maintenance, and technical support. Of course, even to get the plant's attention, we must first have been successful in receiving a positive endorsement from corporate management. Without this endorsement, we may get nothing more than a polite hearing from the plant personnel, if that. Notice, also, that when dealing with the plant, operations and technical support will play a role that could be just as important as the maintenance role in achieving a successful RCM program.

To be more specific, there are three important issues that interplay with one or both of these organization structures:

1. approval and funding

2. buy-in

3. resource allocation (mainly manpower)

Let's look at each issue separately.

Approval and funding.   What kind of funding does it take to properly conduct an RCM program that will generally adhere to the sys-

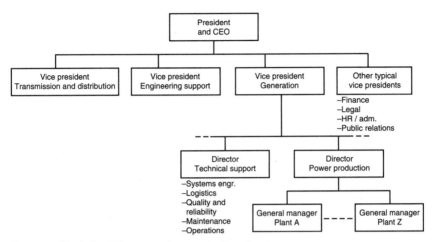

Figure 8.1 Typical utility corporate organization structure.

tems analysis process described in Chap. 5? We really need to give at least two answers to that question—one for fossil power plants and one for nuclear power plants—since our experience tells us that the analysis complexities encountered with nuclear plant systems are, on average, significantly greater than with fossil plant systems. In a fossil plant, with 10 of its 25 systems receiving the RCM process, a typical program cost might be in the $150,000–$300,000 range, and it might take 2 to 3 years to complete. In a nuclear plant with 30 of its 100-plus systems receiving the RCM process, a typical program cost might be in the $1.5 to $2.5 million range and take 3 to 5 years to complete. Of course, there are many variables which influence these cost and schedule figures, such as learning curves, personnel experience, salary levels, number of teams employed, and team size. (It is important to note that, when multiple plants or facilities are involved, a new RCM anal-

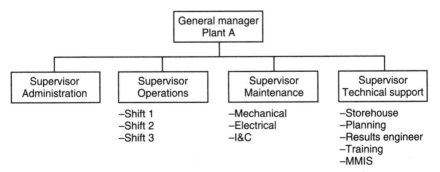

Figure 8.2 Typical utility plant organization structure.

ysis for each system is usually *not* required; rather, the existing systems analyses can be "replicated" at the other plants at a considerably reduced cost. The replication process is a particularly useful concept with fossil power generation units—or any similar situation where a large number of plants/facilities are involved.) But at these levels of expenditure, it is safe to say that the approval requirements are at the corporate level where the vice presidents of engineering and generation, together with key members of their staffs, must formally concur in order to establish a line item in the budget for the RCM program. Thus, the initial sell and buy-in occurs at the corporate vice-presidential level, and this may or may not occur with visible support from the plant general manager. If you are fortunate, it is the plant general manager who initiated the request for RCM program funding because the selling job is then already halfway done at the outset. But this is rarely the case, since plant managers are completely occupied with keeping the plant on-line, and have little time to discover and promote new and innovative ideas without a gentle push from above. In one rather extreme case, the board of directors became the approval authority, but the more general rule is that you must go to the corporate vice presidents to get the ball rolling. If two vice presidents are involved, your selling job may be more than twice as difficult—simply because these people may have different agendas and priorities which tend to compete for available funding. Thus, you may have to gain a very comprehensive understanding of these two agendas, and then find a way to couch your sales pitch to fit both agendas. If one of the corporate vice presidents, per se, is the initiating sponsor, funding approval becomes a virtual certainty, since he or she usually will not take such a visible step without first knowing that success is highly probable. There are also situations where an external force could be the initiator, and one such example has occurred as this book goes to press involving the issue of a "Maintenance Rule" by the Nuclear Regulatory Commission governing certain "shall do" actions for nuclear power plants. In this latter case, certain of these requirements can be satisfied with an RCM program, and several utilities have been motivated to consider RCM as a part of their response to the NRC. Even then, however, the approval and funding decisions usually reside with the corporate-level senior managers.

Buy-in.   In Sec. 8.1.1, we dealt with the subject of buy-in in lesson 2. Here, we want to tie that discussion more directly to the organization structure of Figs. 8.1 and 8.2. The first essential step has already been identified as endorsement of the RCM program from top management. That endorsement, in concrete terms, comes from the corporate senior management approval and funding that was previously discussed. Given this endorsement, the really crucial buy-in actions must now focus on the plant organization in Fig. 8.2. Here, the sales job will (1) be more technical in nature, (2) involve the need to sell a larger group of people, and (3) be more difficult because the "preserve equipment" culture is more firmly established as the daily routine of plant life. Perhaps the most important point to make here, organizationally speaking, is the need to sell technical support and operations personnel as thoroughly as you need to sell the maintenance personnel. The reasons for this are embedded directly in the RCM process itself, where the CD and FF tasks take on equal importance with the more traditional TD (overhaul-type) tasks. And who "owns" a share of responsibility for the CD and FF tasks? Right—operations and technical support. If you should fail to recognize this facet of the plant organization, you may well fail ever to achieve a completely successful implementation of the RCM program—even if the systems analysis process was a success.

Resource allocation.   Where will the personnel to staff the systems analysis process and Task Packaging efforts come from? This has occasionally been a monumental issue which has led to delays in initiating an approved RCM program—delays that in a few instances have literally spanned several years. The nature of the difficulty with this issue involves the unfortunate fact that the most logical place to staff and conduct the RCM program, the plant itself, usually does not have sufficient availability of personnel to do the job. This issue is particularly critical in dealing with RCM programs at fossil plants where on-site staffing is already "lean and mean," but it also can be a critical issue at nuclear plant sites, even though the number of support personnel assigned to the site are relatively generous. The plant, of course, is the logical first choice when we consider the role that buy-in plays in assuring a successful program. There are four possible solutions to the staffing issue, each of which has been employed in the last few years by the utility industry:

1. Bite the bullet and assign appropriate on-site plant personnel to the RCM program by giving it top priority over other activities. The problem with this solution is that the top-priority assignment goes by the wayside every time there is any sort of hiccup in the plant availability status. Personnel are continually pulled away from their RCM team assignments "for just a few days to handle the crisis" with the net effect that a smooth and continuous RCM program really never occurs. Everyone gets frustrated in the process and, in the worst case, the program may never be completed while, at best, the schedule for the program can be extended far beyond the original target dates for implementation.

2. A variation on this theme is to authorize an increase in plant staffing specifically to accomplish the RCM program. With this

approach, some (one or two) key personnel from the existing staff might be placed in lead positions to help orient and integrate the new personnel into the plant community. There have been cases where this approach has worked exceedingly well, and has ultimately produced some of the best implementation results.

3. A third possibility is to staff and conduct the RCM program through the technical support group at corporate headquarters. This is often considered to be the best solution from a staffing point of view, but it also turns out to be a poor solution in terms of the required buy-in at the plant. This latter point can be mitigated to a large degree *if* the corporate-staffed RCM teams play a continuing and highly visible role with involvement from and integration with the plant personnel. Success with this approach is totally dependent upon how well this "if" is handled.

4. A fourth approach is to bring in an outside contractor to execute the entire RCM program. In this case, there is usually some degree of plant management assigned to oversee the contractor, with the net result that a very minor participatory role is played by the plant staff. This condition not only creates a major buy-in problem at the plant, but it may also create some technical deficiencies in the systems analysis process, since the contractor personnel may not have the in-depth systems knowledge required to thoroughly perform the Seven-Step systems analysis process. The results of this approach have been mixed. There have been successful programs, and there have been partial to total wipeouts (i.e., the contractor's product was partially or totally nonusable). This approach is usually the most expensive approach, and has the lowest probability of yielding completely satisfactory results. A variation of the contractor theme has been employed by several companies with whom the author has worked. The examples used in Chap. 7 are illustrative of this theme, where the company has used one of the first three approaches just described, and has employed a *single* consultant to work with the RCM teams until they become RCM process experts in their own right, and can then complete the program using in-house personnel exclusively.

In summary, it is the author's view that approach 2, augmented during the early program phases by an expert outside consultant, offers the highest probability of RCM program success.

8.1.3 Using CD and FF tasks

Throughout Chaps. 5, 6, and 7, it has been shown that CD and FF tasks will constitute a significant percentage of an RCM-based PM program. This situation, previously listed as one of the "Key lessons to remember" in Sec. 8.1.1, introduces a rather unique issue that deserves some specific attention.

Traditionally, a plant PM program is considered to encompass those tasks which are limited almost exclusively to what we have defined here as TD tasks—that is, the overhaul and intrusive type of maintenance actions. Thus, the maintenance department is composed of craft personnel trained in mechanical, electrical, and I&C equipment failure prevention, repair, and restoration skills, and they respond to either the scheduled tasks that have been placed in the MMIS or the requests for corrective actions that most frequently come from the operations department. If any CD or FF tasks are performed, they may be assigned to the operations or technical support department, not the maintenance department, depending on the particular organizational philosophy that is invoked at each plant.

Along comes RCM and, in a sense, it introduces the somewhat foreign notion to many plant personnel that CD and FF tasks, together with TD tasks, now constitute the PM program. When operations personnel are assigned CD and FF task responsibility, they suddenly discover that they have become an integral part of preventive maintenance. Most of the time, the initial reaction to this is shock and disbelief, followed quickly by resistance.

To cut to the core of the issue, RCM brings to the surface the need for operations and maintenance to share mutually in the betterment of the plant maintenance—rather than to continue the age-old adversarial relationship that traditionally exists to one degree or another. RCM *requires* that the operations and maintenance departments combine their respective talents in an integrated PM program effort. Failure to get such cooperation and integration can doom an RCM program to mediocrity or even outright failure. Thus, operations must be part of the buy-in process and must participate actively in defining, and then implementing, the RCM-based PM tasks.

Recall in Chap. 2, Sec. 2.3 that CD tasks come in two forms: (1) performance parameter analysis from *existing* operating instruments and (2) ancillary instrument readings and measurements (usually referred to as *predictive maintenance* techniques). Since the operations department "owns" the operating instruments, they therefore also "own" the ability to develop a variety of CD tasks. Experience has shown that the operations personnel are very astute at using these in-situ instruments to control the plant, to monitor plant status on a periodic go or no-go (attributes only) basis, and to measure certain plant parameters that are critical to peak operating efficiency. This same experience also has shown that the operations personnel can, with little additional effort, convert many of their attribute measurements to trend measurements, and thus create several meaningful CD tasks for the PM program. Reliability-Centered Maintenance is beneficial in uncovering such potential opportunities that have gone unnoticed for years. For example, in Chap. 7, Fig. 7.8, we see the category "RCM task equals current task with some modification." The "some modification" referred to here is frequently the conversion of an existing attribute measurement by operations to a trend measurement over time in order to predict incipient failure conditions. In Fig. 7.8, virtually all of the 37 tasks so identified for the combustion air delivery system were of this nature.

The introduction of CD and FF tasks to a PM program is a blessing—not a punishment. It is necessary that this fact be understood by all parties concerned.

8.2   Computer Usage in RCM

Ah, the computer—what a wonderful device! It does almost everything for us, and life is so much simpler now that is has become a part of our everyday routine. Right? Well, not quite. There certainly can be no argument that the computer has had a profound impact on the world over the past 25 years—an impact that has had some mixed blessings. On the plus side, there are technical breakthroughs that could not have occurred without the ability to handle voluminous computations and data manipulations which could only be performed with the power of a digital computer. Such examples would include space flight (orbital and planetary flight mechanics), modern jet aircraft (computational fluid dynamics and autoflight control), exotic skyscrapers (finite element analysis), nuclear power (atomic fission phenomenon), and the list could go on. In the business world, we also see the vast majority of transactions being handled by the computers. These transactions range all the way from the simplest fast-food purchases to the millions of bank accounts processed daily to the instantaneous handling of travel reservations of all types. But, unfortunately, with all of this "good," we have experienced some "not-so-good" from the computer. Three such areas concern me personally quite a bit: (1) The computer has made our personal affairs somewhat of an open book, since virtually every piece of pertinent statistical data ever recorded about us has been placed in a computer file somewhere where it is accessible by others. (2) As a society, we have become a slave to the computer, relying upon it to substitute for even the simplest of thought processes (are we forgetting how to think?) and blaming it for anything and everything that goes wrong. And (3) we have permitted a proliferation of computers to the extent that almost everyone in the white-collar workforce has a PC on his or her desk, even though only a very small percentage of this population *productively* uses the PC on a regular basis. Maybe you, too, have a pet peeve or two that could be added to the "not so good" list.

What does all of this have to do with RCM? Simply put, the author is continually asked, "When are you going to computerize the RCM process?" Frequently, the intent behind this question is really driving at having the computer "solve" the PM program optimization process automatically by providing a few simple inputs to some magic software code. Well, here is my answer to these queries.

First, there is no magic software code to do the engineering thinking for us in the RCM process. There could be some RCM Expert Systems sometime in the future to help us with our thinking, but it does not exist today. Second, since the RCM process is a qualitative process, there are no voluminous computations and data manipulations to perform. So the computer can't play a role in that arena. That leaves really just one area where the computer can be a valuable asset to an RCM program—namely, as a word processor and data organizer. And in this capacity, the computer can play a real and important role in the RCM process.

Recall from Chaps. 5 and 6 that the Seven-Step systems analysis process uses some specific forms and formats to organize the engineering information in a progressive and logical fashion, and that some of that information is carried from one step to the next for ease of traceability and visibility. If we wish, we can computerize the various forms, enter the information directly on the forms as it is generated, and, where there is information carryover to the next form, we can automatically create that carryover with the touch of a button on the keyboard. Once this information is committed to the computer disk, we can also develop software that will look upon this collection of information as a database, and will allow us to manipulate the data in a variety of useful ways. For example, we can develop a historical file of failure modes for each equipment type that can be used to help in constructing FMEAs for several systems with common components. Another useful file would be a compendium of PM tasks that have been considered for the failure modes. In fact, we could even go to a paperless RCM process where the entire collection of RCM analyses and results for an entire plant could be placed on a couple of floppy disks. For those who like to "save a tree," this latter feature should be especially appealing.

In the early RCM pilot projects in the utility industry (Refs. 13, 14, and 15), computers were conspicuous by their absence. However, when nuclear plants first became serious about doing 20 or 30 systems with the RCM process, the computer started to emerge, first as a working word processor to completely bypass handwritten forms that needed to be placed later on the computer file, and then with a combined word processor and database capability. Two of the earliest computerized capabilities were the successful in-house efforts of Florida P&L Company, with its fossil RCM program, and GPU Nuclear Corporation, with its TMI-1 RCM program. Later, the Electric Power Research Institute extended its RCM efforts to include the development of an

RCM workstation software specification (Ref. 24), and then the RCM workstation itself which was developed by Halliburton NUS (Refs. 25 and 26). In the meantime, other organizations (including Westinghouse, ERIN Engineering, Quadrex, and Life Cycle Engineering) continued to develop their own RCM computer capabilities.

Thus, the computer has, in fact, found its place in the sun in the RCM process. It doesn't replace the need for solid engineering know-how and judgment, and it doesn't compute anything. But it can do a very effective secretarial and data retention job for you—in its own automated way.

8.3   Task Periodicity and Age Exploration

The RCM process is the methodology for defining "what task," but it does not directly define "when done." We must therefore separately consider the process that will lead to the selection of a periodicity (or interval or frequency) for each task in the RCM-based PM program.

If the task selection process has employed the road map presented in Chap. 5, Fig. 5.18, we will have at least established at the outset if we know the age-reliability relationship for the specific failure mode in question. This relationship, as we have seen, is the key item of information that is needed to initially consider the practicality of seeking a TD task whose objective would be to prevent the onset of a known aging or wearout failure mechanism. Now, if we do know the age-reliability relationship, then we also have the precise information that we need to select the TD task interval. That is, we have the failure density function (fdf) for the failure mode population, and we can select the task interval from this statistical knowledge by simply deciding on the level of consumer risk that we want to accept. Suppose, for example, that the fdf looks like the bell-shaped curve shown in Chap. 3, Fig. 3.1, where the x-axis is operating time and the y-axis is probability of failure. The left-hand tail may be quite long, thus signifying an extended period of time during which the probability of failure is quite small and, for all practical purposes, is in a constant failure rate condition. That is, the aging/wearout failure mode has not yet really started to exhibit itself. Recall that we did indeed see such situations in the curves shown in Chap. 4, Fig. 4.1 (curves D, E, and F). However, as we proceed to the right in Fig. 4.1, or as we see the probability of failure beginning to increase as additional operating time is accumulated, we can now decide just how far we want to proceed before doing the TD task. And this is where the level of consumer risk comes into play. We can pick that level of risk by selecting the percentage of area under the fdf that we can tolerate before taking action. Say we choose 15 percent. This means that there is 15 percent chance that the failure mode could occur before we take the preventive actions. Notice that we can choose any percentage value that we want; the only question is how much risk do we want to take. Notice that if we use the mean (or MTBF) for the bell-shaped fdf, there is a 50 percent chance of failure before we take preventive actions. For other fdfs, the chance of failure can be as large as 67 percent when the mean is used. This is clearly not an acceptable level of risk in most circumstances—hence, using an MTBF value is not really a valid and useful technique for selecting task intervals.

The foregoing discussion has briefly outlined the most ideal situation that we can experience for selecting task intervals. This ideal is not encountered as often as we would like to see because we usually do not have sufficient data from operating experience to define the fdf. So let's discuss what we can do in the nonideal situations that are more commonly encountered. The first situation is one wherein we have a partial knowledge of the age-reliability relationship. This means that we have some operating experience to support the conclusion that aging/wearout mechanisms exist, but we do not have any statistical data to speak of to define an fdf. So we tend to use our experience to guess at a task interval for the TD actions. In so doing, there is overwhelming evidence to show that this process is highly conservative. That is, we tend to pick intervals that are way too short. We might overhaul a large electric motor every 3 years when, in reality, the correct interval turns out to be 10 years! We must learn to correct this conservatism because it is costing us dearly. We do so via Age Exploration which is described subsequently.

Again referring to Chap. 5, Fig. 5.18, the second situation is one in which we have no idea what the age-reliability relationship might be, and we are now moving on to look for candidate CD tasks. If the failure mode is hidden, we also extend our search to include candidate FF tasks. These tasks, too, must have intervals specified for the nonintrusive data acquisition and inspection actions that must be accomplished. And, here again, the statistical basis for specifying these intervals is usually missing, and we guess at what they will be—and usually with great conservatism. So Age Exploration will be useful to us with CD and FF tasks as well as with TD tasks. Let me offer one cautionary note about CD tasks. When we select a CD task, we must specify not only the task interval, but also the parameter value that must be used to alert the plant personnel that the incipient failure process has begun. Selecting the correct value may also be a guessing process at first, and additional experience must be systematically collected to adjust that value over time so that the alert is neither too early nor too late.

When good statistical data is not available, using our experience to guess at task intervals is really the only option that is available to us initially. But there is a proven technique that we can employ to refine that "guesstimate" over time, and to predict more accurately the correct task interval. It is called *Age Exploration,* or AE. The AE technique is strictly empirical, and works like this (using a TD task for illustrative purposes). Say our initial overhaul interval for a fan motor is 3 years. When we do the first overhaul, we meticulously inspect *and record* the condition of the motor and all of its parts and assemblies where aging and wearout are thought to be possible. If our inspection reveals no such wearout or aging signs, when the next fan motor comes due for overhaul, we automatically increase the interval by 10 percent (or more), and repeat the process, continuing until, on one of the overhauls, we see the incipient signs of wearout or aging. At this point, we stop the AE process, perhaps back off by 10 percent, and define this as our final task interval.

Figure 8.3 illustrates how this AE process was successfully used by United Airlines for one of their hydraulic pumps. On the top half of Fig. 8.3, we see that the overhaul interval started at about 6000 hours, and that the AE process was then employed over a four-year period to extend the interval to 14,000 hours! The bottom half of Fig. 8.3 presents a second very interesting statistic for the same population of pumps over the same four-year interval. The statistic is *premature removal rate* (or the rate at which corrective maintenance actions were required). The interesting point here is that the premature removal rate has a definite decreasing value over the four-year period where the overhaul interval was increasing. We interpret this to suggest that as the amount of human handling and intrusive overhaul maintenance actions decreased, so did the human error resulting from such actions, with the net effect that corrective maintenance actions likewise decreased. Recall that we saw this same situation in the statistics presented in Chap. 2, Fig. 2.2.

While it is certainly true that Age Exploration can be a lengthy process, one should consider that it is really the best alternative available when the statistical process cannot be used. As a rule, the collection of large samples of statistical data can be an even longer process.

8.4   The Living RCM Program

The systems analysis process in Chap. 5 is a one-shot effort and, when completed, it represents a baseline definition of the PM program for the system in question. However, we need to recognize four factors that suggest the need for some continuing RCM program activity in order to continuously harvest the full potential of the RCM process:

1. The systems analysis process is not perfect, and may require periodic adjustments.

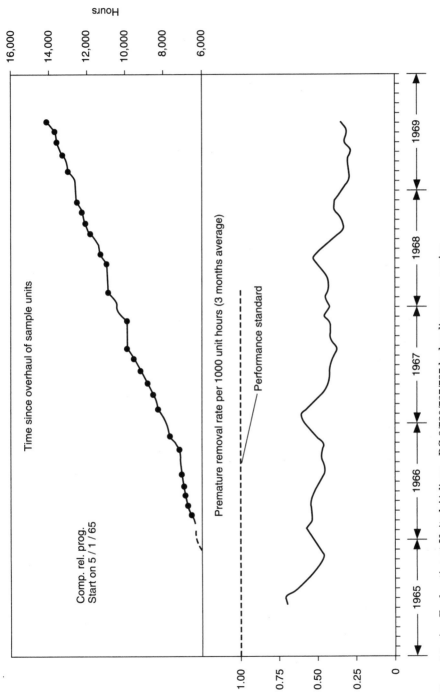

Figure 8.3 Age Exploration at United Airlines—DC-8/720/727/737 hydraulic pump—engine driven.

2. The plant itself is not a constant since design, equipment, and operating procedures may change over time.

3. Knowledge grows, both in terms of our understanding of how the plant equipment behaves and how new technology can increase availability and reduce O&M costs.

4. We need to measure the cost benefits derived from RCM.

The continuing activity suggested by these four factors is what we call "The Living RCM Program." We will briefly discuss each of the four factors.

Adjustments to the baseline results.   The main consideration here revolves around the question of whether the Baseline Definition is totally correct. The most likely answer is "no." But how do we know this? The best way to understand this is to periodically review the *unexpected* corrective maintenance actions that have occurred (which does *not* include such actions for RTF decisions). This measure will provide a direct reading of where the Baseline Definition was either in error or missed something important. If we are experiencing unexpected failure modes in spite of our PM actions, then the original PM action selection may be the wrong thing to do and some adjustment may be in order. If the unexpected failure can be traced to a failure mode that is not covered in the Baseline Definition, then we need to include it as an amendment to the original systems analysis process, and reevaluate the necessity for including a new PM task.

Plant modifications.   Even though a plant or facility is literally cast in concrete, it is rare indeed to encounter a situation where there are no changes to the plant over its operating lifetime. These changes occur for a variety of reasons—such as capacity enlargement, productivity improvements, safety and environmental enhancements, and regulatory enforcements. They may involve new additions, redesign of existing systems, replacement of components with upgraded features, and alterations to operating procedures to reduce equipment stress or increase efficiency. Any such change should be reviewed against the PM baseline definition to ascertain if new or modified PM tasks are needed and, in some instances, to delete PM tasks that are no longer applicable and effective.

New information.   Our knowledge base is continually increasing. We learn about the "personality of the plant" as our operating experience grows, and (hopefully) we collect operating data in the MMIS which expands our ability to analyze and understand the equipment behavior. This expanded knowledge of the plant behavior may tell us that the PM program requires some adjustments. For example, our knowledge acquired from the Age Exploration process permits us to adjust task intervals. We must also recognize that predictive maintenance technology is expanding. New techniques for condition-directed tasks are emerging as you read this book. Thus, we may find that PM task effectiveness can be increased with this new knowledge *if* we use it to our advantage.

Program measurement. Even if the baseline definition never changes, at a minimum we should measure the benefits derived from the RCM program as a part of the routine plant operating reports to management. Of course, management will be particularly interested in how RCM has impacted the bottom line. Changes to the baseline definition should also be measured to assure that the PM task effectiveness criteria have, in fact, been optimized. Suffice it to say that such measurements are a tricky process—for example, they can be so global in nature that it becomes very difficult to sort out the parameters that are governing the observed result; on the other hand, they could be so abstract that it is impossible to define clearly a meaningful message. Plant availability or capacity factor is a typical global measurement; it is a very important measurement, but so many factors can influence its rise or fall that it may be next to impossible to pinpoint the precise reasons for a change. The PM to CM cost ratio is, in the author's view, a very abstract measurement, and one that has neither good nor bad values. For example, in a well-constructed RCM program, we have seen that RTF decisions are an important part of the total makeup. How does one account for the influence of RTF decisions on the PM:CM cost ratio in deciding what ratio values are good or bad? Given the preceding caveats, there are three measurements that historically have proven to be useful at a minimum:

1. *Unexpected failures.* As noted previously, this measurement is very valuable in fine-tuning the PM baseline definition for each system. Over time, the occurrence of unexpected failures should approach zero.

2. *Plant availability.* Even though this is a global measure, it does fairly represent a very important indicator of plant performance. And, as plant availability increases, *cost avoidance* accruals can be a major bottom-line benefit (i.e., avoidance of costs or income losses associated with plant downtime).

3. *PM + CM costs.* This *total* cost figure, tracked over time, gives an excellent measure of just how the RCM program is affecting maintenance expenses. It is the total that counts, not the individual values. If RCM is doing its job, this total should decrease over time.

The final question to consider is the frequency with which the Living Program formal review should be conducted on the baseline definition of each system. Please recall that the accumulation of information for each of the previously discussed four factors is, itself, an on-going and continuous process. The question here, then, is directed at how often we need to take this information and specifically compare it to the existing RCM program documentation. To a large degree, the answer here is strictly a judgment call. In the case of a major unexpected failure, a major plant modification, and the like, an immediate review may be in order. However, it is more likely that the formal review should be conducted every 12 to 24 months. With this interval length, the resources required for the Living Program are fairly minimal, and we have allowed sufficient time for items requiring adjustment to appear. In all likelihood, the need for adjustments will diminish in time, and the frequency of the Living Program reviews will increase to the 36-month range.

Well, there you have it. Hopefully, you are now an RCM expert, and know how to go about optimizing your PM program. Good luck—and go to it!

The Preventive Maintenance Role in Total Quality Management

A.1 The Challenge—The Need

In the first 25 years after World War II, the demand for U.S. products, technology, and services was virtually unending. To be sure, demand outstripped supply, and U.S. business flourished at home and on the world market with unprecedented levels of success. In many respects, we enjoyed a virtual monopoly in the international arena. Our success stemmed from customer-driven inputs and a work ethic that motivated all of us to do our very best at our jobs. With only a handful of exceptions, our households and factories were composed of products "Made in the USA." (One interesting exception that swept the nation, but never really posed any potential economic calamity, was the Volkswagen "Beetle.")

We all know what happened in the 1970s and 1980s! We woke up one day to find that we had some unbelievably good competition from abroad—especially Japan. It was not only gobbling up a large percentage of market share in crucial industries (steel, autos, ship building, etc.), but this competition was actually driving U.S. industry right out of certain product lines—such as TVs, VCRs and other consumer electronics. Again, the Japanese were the major culprits, with the Europeans coming on strong in several markets. They had all done their homework, and done it well.

Some other things happened in the 1970s and 1980s that rocked our economic chairs fairly violently. For example, OPEC sent energy into a crisis state, inflation and interest rates went berserk for extended periods, and *American product quality started to take a back seat to foreign quality. Clearly, this latter development was the basic root cause of our difficulties*—and to a major extent, continues today to be a basic problem in U.S. industry. As we look forward to the next decade, the scene does not necessarily look rosier. The foreign competition is growing (e.g., the introduction of the European Community in 1992 and the emergence of the eastern bloc and third world countries) and, by all accounts, we have a great deal more to do to get our collective act together.

Where did we go wrong! Are our foreign competitors really better than us? Have we lost the technology edge? Has management lost its touch? Has labor lost its craftsmanship? We could probably answer at least a partial "yes" to each of these important questions. But most importantly, what do we do about it? This question has been studied and restudied in recent years by very learned scholars, high-level government task force committees and any number of large U.S. companies. The answer that has consistently come forth from all such sources has had one overwhelming common thread. *We must improve quality*—product quality, service quality, personal commitment to quality in our daily lives.

That is our challenge; that is our need. It is now a well-established fact that *any* country or corporation that wishes to achieve a long-term market share in the global arena *must* improve its quality stature and recognition—or fade from view forever.

There are many proven ways to approach quality improvement. The most successful of these, as espoused by Deming, Juran, Crosby, Feigenbaum, and Harrington, are whole stories in their own right. We do not intend to develop these stories here in any detail. What we do intend to do is show how the basic structure of a Quality Improvement Program (QIP) can be developed and, in turn, how such a structure can lead to a very definitive set of actions that will delineate an optimization of preventive maintenance programs where large and complex plants and/or systems are an integral part of the ultimate product (or its development and manufacture). We do this in order to emphasize the need for excellence in the conduct of maintenance actions as well as the logical and important role that preventive maintenance can play in a Total Quality Management (TQM) approach to continual improvement of products and processes.

A.2 Total Quality Management (TQM)

"Quality" has been in our vocabulary for a long time. Technically speaking, quality assurance and quality control programs have been an essential and formalized element in product development and manufacturing cycles for several decades. In the technical context, *quality* has come to mean conformance to specifications, processes, and drawings. There is nothing inherently wrong with this meaning—except that this form of quality could insist, for example, on the continued output of products that are flawed in their basic design or makeup. To the man on the street—that is, the customer—quality has a much broader and less precise meaning. In the nonprofessional context, quality relates more to customer satisfaction, and this can introduce a variety of parameters that range from an objective technical performance measurement to some subjective feeling of goodness about a product's color, shape, or feel.

When we talk about improving quality, just what do we mean? What quality are we talking about? In recent years, this question has been answered by speaking about *total quality.* Simply put, *total quality* is meeting the desires, needs, and expectations of customers.

Notice that total quality, the quality where improvement is really needed, is the one that relates directly to the nonprofessional context and meaning. Oddly enough, the total quality concept is bringing us full circle back to where we were in the 1950s and 1960s!

Thus, *Total Quality Management,* or TQM, is the establishment and execution of the system and corporate culture that will ensure customer satisfaction in the products and services delivered.

The definitions of total quality and TQM are simple enough, and their objectives certainly make sense. But how do we do it? Basically, TQM is based on four principles:

1. *Customer Satisfaction.*   Quality is satisfying the customer. Satisfying the customer means meeting their needs and reasonable expectations. Beyond that it means having an attitude that puts the customer first. (For example, a phone call from someone who uses one of my products is not an interruption from my work. It *is* my work.)

2. *PDCA.*   Plan-Do-Check-Act, sometimes known as the Deming circle. This is a four-phase philosophy for working and problem solving that is embedded throughout any organization espousing TQM:

    Plan what to do.
    Do it.
    Check results.
    Act to prevent future error or to improve the process.

3. *Management by fact.*   (Often referred to as "speaking with facts.") This has two meanings not only for managers, but for all employees: First, collect objective data. Second, manage according to this data.

4. *Respect for people.*   This principle assumes that all employees have a capacity for self-motivation and for creative thought. Each employee needs to listen to, and support, this capacity in every other employee.

The TQM principles are being embraced by a large segment of U.S. industry. Implementation, however, is taking various paths. For example, in the automobile industry we have seen each of the "big three" develop a single product line along TQM lines as a trial to model the entire company (Chrysler—the K car, Ford—the Taurus and, most recently, GM—the Saturn). With the Department of Defense, the Pentagon has decreed that all major procurements will require the selected contractor to use TQM. This has resulted in a rash of TQM programs in major aerospace firms, and in one instance (McDonnell Douglas) even resulted in a corporatewide major reorganization to structure for a total quality orientation.

One of the earlier and very successful TQM programs undertaken within U.S. industry is that developed by Florida Power and Light Company (FPL). Initially, FPL developed its own Quality Improvement Program, or what came to be called *QIP*. QIP invokes the four basic principles of TQM, and additionally has established a subset of key principles:

1. QIP must be how we manage our business.

2. Eighty percent of problems are under management control.

3. A small number of problems have the most impact.

4. People doing the work can solve problems.

5. Process must be stressed.

6. Corporate systems and structures must support desired behavior.

7. Reduced cost, improved reliability, and increased productivity are natural consequences of improved quality.

This program, as implemented via the QIP triangle that is depicted in Fig. A.1, was the approach employed by FPL in their successful effort to be the first U.S. company to win the Japanese "Deming Award" for total quality excellence.

In the next section, we will briefly summarize the Quality Improvement Program at FPL in order to illustrate briefly one successful TQM approach. We will then use the QIP policy deployment process in Sec. A.4 to illustrate just how TQM has identified the importance of preventive maintenance in the total quality picture.

A.3   The FPL Quality Improvement Program (QIP)

A.3.1   Reasons for QIP

In the period from 1946 to 1974, FPL experienced rapid growth. The sales growth rate was in double-digit figures through 1973 and it was difficult for the company to keep up with the need to plan, finance, construct, and operate an electric system to serve a growing south Florida. In addition to expanding its service, FPL maintained stable prices for its customers. Then came the first oil crisis and a resulting period of high inflation. Sales growth slowed, the company's stock price fell, and bond rates went up. In 1978, the government passed the National Energy Act, which resulted in competition for utilities and promotion of conservation.

Due to these conditions, the trust the company had developed with its customers suffered greatly. The news media highlighted issues of cost, reliability, and safety. The government increased regulation over nuclear power plants as a result of the Three Mile Island accident in Pennsylvania in 1979. The increased government regulation, combined with public concern for nuclear safety, led to the cancellation of many plants under construction and extended shutdowns for many of those in operation. This increased the uncertainty of the future of the nuclear power supply. Regulation in other areas, such as the environment, also increased. Rates continued to rise, placing political pressures on regulators that resulted in a breakdown in the relationship between regulators and utilities. The price increases were neither timely nor popular because customers also were feeling the effects of inflation.

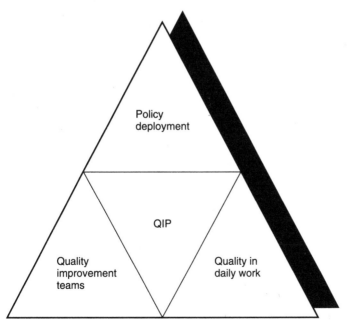

Figure A.1   Florida Power & Light's quality improvement program triangle. (*Courtesy of Florida Power & Light.*)

By the early 1980s, FPL was facing a hostile environment created largely by high inflation, decreasing customer sales, rising electric rates and increasing fuel oil prices. This environment caused the price of electricity to increase faster than the Consumer Price Index (CPI). The company began to realize the significance of competition. Also, the company recognized a decline in customer satisfaction due to increased expectations in the areas of reliability, safety, and levels of customer service.

Working in this external environment, FPL's top management also began to identify and address some important internal issues impacting the effective management of the corporation. For example:

1. As FPL grew, it had become more bureaucratic, and meeting the needs of the customer in a timely manner became difficult.

2. Management could foresee no significant technological innovations to reduce the escalating power-supply costs. Therefore, a change in management philosophy was needed to achieve customer satisfaction and cost reduction through greater management effectiveness.

Management concluded that the company's internal and external environments were changing faster than the organization could adapt, and that corporate goals needed to be established and achieved using new management techniques.

The identification of selected significant issues that were impacting company survival are further illustrated in Fig. A.2, and the resulting corporate goals to address these issues are shown in Fig. A.3.

A.3.2   Development of QIP

FPL had to change its way of thinking from supply-oriented to customer-oriented, from a power generation company to a customer service company. This new strategy needed to provide a means to address the key issues surrounding the satisfaction of customer needs and expectations.

Quality efforts began in 1981 with a limited and narrow approach called *quality improvement (QI) teams,* similar to Japan's quality circles. Management soon realized that QI teams alone would not achieve the results needed to change the company.

In 1985, FPL chairman of the board and chief executive officer, John J. Hudiburg, announced that the company would fully implement its Quality Improvement Program (QIP). Mr. Hudiburg told FPL employees that after thoroughly investigating TQM methods, he was convinced that TQM was the appropriate management system for FPL. He also told them that the QIP philosophy and system would be deployed throughout FPL and that QIP would be the method used to accomplish the corporate vision.

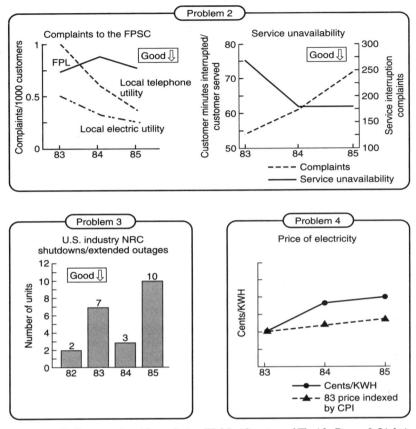

Figure A.2 Indicators of problems facing FP&L. (*Courtesy of Florida Power & Light.*)

FPL's QIP consists of three major components, shown in Fig. A.1:

- *Policy deployment.* Policy Deployment is the process in which company management works together to focus resources on achieving customer satisfaction.

- *Quality in daily (repetitive) work (QIDW).* QIDW provides a decentralized method for controlling and improving daily work processes. Its basic aim is to ensure that routine activities are performed correctly. Also, it provides for maintaining the gains achieved through improvement activities that employ the plan-do-check-act (PDCA) cycle.

- *Quality improvement teams.* QI teams were developed to provide a structure for the employees to improve the quality of products and services, to develop their skills and abilities, to promote better communication and teamwork, and to enhance the quality of their work life. It gives all employees the opportunity to be heard by management and to express their individual creativity.

```
┌─────────────────────────────────────────────────────────────────────┐
│                         FPL Corporate Vision                         │
│     During the next decade, we want to become the best managed       │
│     electric utility in the United States and an excellent company   │
│     overall and be recognized as such.                               │
└─────────────────────────────────────────────────────────────────────┘
                                    ▲
                                    │
┌─────────────────────────────────────────────────────────────────────┐
│                            Company Goals                             │
│                                                                       │
│   1. Improve customer satisfaction with sales and service quality:   │
│         (a) by reducing customer complaints to the FPSC to be among  │
│             the lowest in the electric utility industry,             │
│         (b) by having adequate capacity to meet the needs of         │
│             existing and future customers and                        │
│         (c) by having service unavailability among the lowest in     │
│             the industry.                                            │
│   2. Strengthen effectiveness in nuclear plant operation and         │
│      regulatory performance:                                         │
│         (a) by achieving nuclear plant availability to be among the  │
│             highest in the industry and                              │
│         (b) by improving nuclear safety through the achievement of   │
│             reduced automatic plant shutdowns (auto trips/scrams)     │
│             and NRC violations to among the lowest in the industry.  │
│   3. Improve utilization of resources to stabilize costs:            │
│         (a) by improving quality,                                    │
│         (b) by maintaining stable and reasonable prices, as compared │
│             to the Consumer Price Index (CPI), while maintaining a   │
│             fair rate of return for stockholders and                 │
│         (c) by securing the safety of our employees and community.   │
└─────────────────────────────────────────────────────────────────────┘
```

Figure A.3   Corporate goal setting. (*Courtesy of Florida Power & Light.*)

FPL's primary motivation in introducing QIP was to establish a management system and corporate culture to assure customer satisfaction. A fundamental change would be to listen to the "voice of the customer" and to identify their needs and expectations. As a result, a number of short-term plans (STP) were initiated to address priority areas.

A.3.3   Results from QIP

The results obtained from the implementation of QIP at FPL have been dramatic. A series of indicators have been established and tracked in order to measure company progress against the goals enumerated in Fig. A.3. The results through mid-1989 are shown in Fig. A.4 and A.5. In every instance, the key indicators have made a significant move in the favorable (good) direction.

The successful implementation of QIP has allowed FPL to evolve its quality program. This evolution has been made possible by the understanding of and commitment to the underlying principles and techniques of quality improvement by its employees. The change is felt to provide the flexibility necessary to respond to today's challenges and provide FPL's customers with the highest-quality service.

A.4   Preventive Maintenance and Policy Deployment

Policy deployment can be defined as the management process for prioritizing and focusing company resources on the goals and activities at

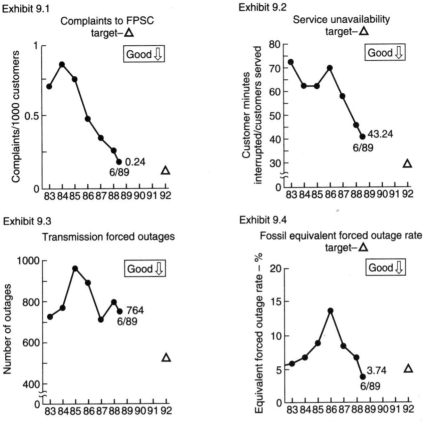

Figure A.4   QIP results versus goal. (*Courtesy of Florida Power & Light.*)

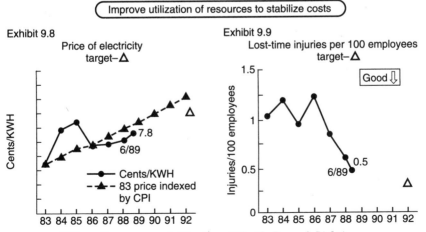

Figure A.5   QIP results versus goal. (*Courtesy of Florida Power & Light.*)

all levels that support the corporate vision. One might consider the policy deployment process to be a form of strategic planning—but the focus is on corporate *improvement* (not corporate direction) that must be realized to fulfill *customer needs* and expectations (not company needs). In this section, we will develop an example of how TQM, and the policy deployment process in particular, will lead to a need for optimizing a preventive maintenance program.

For our example, we will talk about the ABC Corporation, which produces widgets in several plant locations for distribution throughout the United States and several foreign countries. In general, these plants are fairly complex, high-volume facilities, and their productivity is a major factor in the financial viability of ABC. While the widgets continue to be in high demand, the competition for market share is keen.

In applying TQM principals, ABC has done some extensive market research and customer surveys to ascertain the quality elements that are vital to meeting customer needs and expectations. Two such elements that are high on the list are as follows:

1. *Supply*—can deliver the needed quantity on schedule.

2. *Price*—delivered cost per unit is reasonable.

Thus, ABC management must devise and disseminate the corporate policies that will focus the resources on these quality elements. We will construct an example of the policy deployment process that could be used to meet quality element 1.

Phase	Organization source	Policy	How achieved (means)
Quality element	Customer	Deliver ordered items on schedule	
Top management policy	Corporate headquarters CEO and staff	Maintain output	Control plant downtime
Specific goals	Product division GM	Minimize forced outages (F.O.)	Improve F.O. rate
Specific actions	Plant manager	Reduce F.O. events	Develop F.O. countermeasures
Specific tasks	Plant staff		Availability improvement program 1. _____ 2. _____ 3. Optimize preventive maintenance program 4. _____

In this simple but realistic example, we see how the TQM principal of customer satisfaction has driven the ABC management to identify a specific top-level policy (maintain output) and the chief issue to be addressed (control plant downtime). At successive levels in the organization, management has had to translate that policy and issue into the next level of detail that is needed to achieve the desired result. Finally, at the plant level, we see specific actions and tasks unfold which will form the near-term activities for focus of resources. Specifically, the plant has formulated an *availability improvement program* which includes the need to upgrade and optimize the preventive maintenance program.

Clearly, any corporation that requires the use of complex plants and facilities in order to produce its customer's product could develop the above policy deployment scenario and conclude that preventive maintenance plays a key role in their TQM approach.

A.5 Preventive Maintenance Role in Availability

When the ABC Corporation management formulated its top-level policy to "maintain output," they were really talking about a plant characteristic known as *availability*. In simple terms, availability is a measure of the percentage (or fraction) of time that a plant is capable of producing its end product at some minimum acceptable level. For example, an ABC widget plant may have to produce 50 widgets per hour on a normal two-shift day to meet the committed delivery quantities and schedule. If it cannot do this, either overtime shifts may be necessary (if the plant can run at all) or an inventory must be maintained to allow for grace periods in returning an incapacitated plant to service. Usually, such workaround procedures are costly, and it is much more cost-effective to do the things that are necessary to keep the average plant throughput at 50 widgets per hour (i.e., to keep the plant availability as close to a specified goal as possible).

Clearly, forced outages (i.e., inability to maintain a throughput of 50 widgets per hour due to an unplanned equipment stoppage or failure) are the major culprits that reduce plant availability.* Hence, an availability improvement program will consist of tasks that are aimed at avoiding these forced outages. But how is this really done? To answer this question, we need to understand more precisely just what consti-

---

* Availability, by definition, considers all outage activity—both planned (scheduled) and unplanned (forced). Scheduled outages are factored into production commitments as a matter of course, as are forced outages at some small and acceptable level. Big problems, however, can develop when the forced outage rate exceeds the allowable rate. Availability falls below the specified goal.

tutes the availability measure. There are two, and only two, parameters that control this measure:

- *Mean time between failure,* or MTBF, which is a measure of how long, on average, a plant (or an individual item of equipment) will perform as specified before an unplanned failure will occur.

- *Mean time to restore,* or MTTR, which is a measure of how long, on average, it will take to bring the plant or equipment item back to normal serviceability when it does fail.

MTBF, then, is a measure of the plant or equipment reliability (R) and MTTR is a measure of its maintainability (M). Mathematically, we can define availability (A) as follows:

$$A = \frac{\text{MTBF}}{\text{MTBF} + \text{MTTR}}$$

Notice that if the MTBF is very large with respect to MTTR—that is, if we have a very high plant reliability—availability will also be high simply because the MTBF parameter dominates what is physically occurring. Conversely, a very small MTTR can also yield a high availability because even if the equipment fails frequently, it can be restored to service very quickly. Usually, neither of these two limiting cases exist, and we have to work diligently at retaining or improving both the MTBF and MTTR parameters in order to achieve a high degree of plant availability.

Recognizing that MTBF (or reliability) and MTTR (or maintainability) are the parameters that we must influence will now simplify our job considerably. There are several tasks that can be performed, and usually some investigation and evaluation of plant problems and operating practices will reveal where the resources should be focused. In particular, however, the role that an effective preventive maintenance program (PMP) can play in achieving desired levels of availability is of special note. This is true because the PMP can beneficially impact both reliability and maintainability when it is properly specified and conducted. The right preventive maintenance (PM) tasks can, for example, be the primary factor in keeping an item of equipment in top running order—tasks as simple as lubrication and alignment checks performed at specified intervals can be the necessary link to retaining an inherent design reliability. In like manner, the right PM tasks could play a major role in decreasing MTTR simply by the use of periodic on-condition monitoring that would detect failure onset and permit an opportunity for repair or replacement at the timing of your choice, thus avoiding the forced outage.

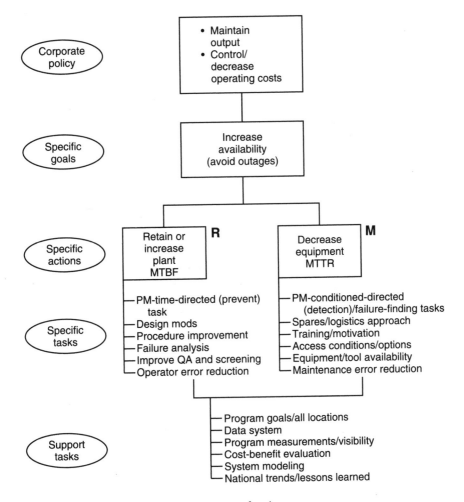

Figure A.6   Availability improvement program key issues.

We can summarize our discussion in Secs. A.4 and A.5 for the ABC Corporation in a single picture shown in Fig. A.6. We can see the basic elements of the policy deployment process in our TQM/QIP approach as well as the essence of the policy–means relationships that were developed. Notice also the other tasks in Fig. A.6 that the ABC management could consider as elements of its availability improvement program. The PMP is particularly potent because, when done properly, it produces a double-barreled effect by impacting both reliability and maintainability within the same program execution.

The Mathematics
of Basic Reliability Theory

B.1 Introduction

In Chap. 3, we discussed some fundamental notions associated with the reliability discipline, and noted in particular the probabilistic or chance element that is basic to its understanding. We also discussed, in very simple and qualitative terms, the mathematical aspects of probability and how this is employed to derive some key elements of reliability theory which are germane to RCM.

In this appendix, we will discuss the derivation of the key elements in reliability theory in mathematical terminology. This discussion is still kept relatively simple, but a cursory knowledge on the part of the reader of some basic differential and integral calculus and probability theory will be helpful to his or her understanding of the process.

B.2 Derivation of Reliability Functions

In Chap. 3, *reliability* was defined as the probability that an item will survive (perform satisfactorily) until some specified time of interest (t). We can think of reliability, then, as the expected fraction of an original population of items which survives to this time (t). Notice that the number surviving can never increase as time increases; thus, reliability must decrease with increasing time *unless* something can be done to essentially return or restore the population items to their original state which existed at $t = 0$. This, of course, is one essential aspect of preventive maintenance actions.

In deriving the reliability functions, we will consider a large number of like items on test and being run to failure. We can thus define the following parameters of interest.

Let N_o = Original population size which will be put into operation at $t = 0$. N_o *is a constant,* a fixed number for the population at t_o.

N_s = Population items surviving at t_x. N_s is a function of time.

N_f = Population items failed at t_x. N_f is a function of time.

$R(t)$ = Reliability of the population as a function of time.

$Q(t)$ = Unreliability of the population as a function of time.

So, at any time, t_x,

$$N_o = N_s + N_f \tag{B.1}$$

$$R(t) = N_s/N_o = N_s/N_s + N_f \tag{B.2}$$

$$Q(t) = N_f/N_o = N_f/N_s + N_f \tag{B.3}$$

And
$$R + Q = N_s/N_o + N_f/N_o = N_o/N_o = I \tag{B.4}$$

That is, R and Q are complementary events; the population items have either survived or failed at time, t_x.

$$R(t) = 1 - Q(t) \tag{B.5}$$

Since N_s must decrease with increasing time, N_f must increase.

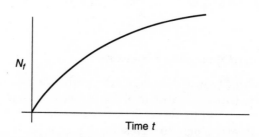

This simple plot depicts the cumulative failure history of our population over time. We can divide N_f by N_o (recall N_o = constant) and the basic shape of the curve does not change. Furthermore,

$$N_f/N_o = Q(t) \tag{B.6}$$

So we now have a curve of Q versus t.

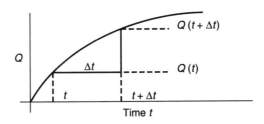

In probabilistic terminology, the curve is the Cumulative Density Function, or CDF, of the population failure history. Also, the derivative of a CDF is a population density function, pdf—or, in this case, a failure density function, fdf.

Taking the derivative of the CDF yields the following:

$$\lim_{\Delta t \to 0} \frac{Q(t + \Delta t) - Q(t)}{\Delta t} = \frac{dQ}{dt} = \frac{d}{dt}\left(\frac{N_f}{N_o}\right)$$

$$= \frac{1}{N_o}\frac{dN_f}{dt}$$

Thus, we know that $\dfrac{dQ}{dt}$ is a failure density function, and let $\dfrac{dQ}{dt} = f(t)$.

$$\text{So, } f(t) = \frac{1}{N_o}\frac{dN_f}{dt} \tag{B.7}$$

From Eq. B.7, we see that in some time interval, Δt, there will be some fraction of the total failures, ΔN_f, that will occur. These failures, of course, are from the original fixed population, N_o, and $\Delta N_f / \Delta t$ represents the total failure frequency in Δt. When this is divided by N_o, the resulting value represents the failure frequency per item in Δt with respect to the *original* population. We call this value the *death rate*.

Thus,
$$f(t) = \text{death rate} = \frac{1}{N_o}\frac{dN_f}{dt} \tag{B.8}$$

A simple example of the death rate calculation was described in Chap. 3, Sec. 3.4.

From here, we can perform various manipulations with the preceding equations to obtain some additional reliability functions of interest.

Taking the derivation of Eq. B.5 yields:

$$\frac{dQ}{dt} = -\frac{dR}{dt} \tag{B.9}$$

Since $\dfrac{dQ}{dt} = f(t) = \dfrac{1}{N_o}\dfrac{dN_f}{dt}$ from Eq. B.7, we know:

$$-\frac{dR}{dt} = \frac{1}{N_o}\frac{dN_f}{dt} \qquad (B.10)$$

Multiplying both sides by N_o/N_s yields:

$$-\frac{1}{R}\frac{dR}{dt} = \frac{1}{N_s}\frac{dN_f}{dt} \qquad (B.11)$$

From Eq. (B.11), we see a similarity with Eq. B.7, the death rate. But now, $\Delta N_f/\Delta t$ is divided by N_s, and the resulting value represents the failure frequency per item in Δt with respect to the population *surviving* at the beginning of the interval, Δt. We call this value the *mortality* or *failure rate*, and assign the symbol $h(t)$ or λ to it. Chapter 3, Sec. 3.4 also gave a simple example of the failure rate and how it is distinguished from the death rate.

Thus,
$$h(t) = \lambda = \frac{1}{N_s}\frac{dN_f}{dt} \qquad (B.12)$$

Also,
$$h(t) = \lambda = -\frac{1}{R}\frac{dR}{dt} \qquad (B.13)$$

Rearranging,
$$\lambda dt = -\frac{dR}{R}$$

Note that at $t = 0, R = 1$.

Integrating,
$$\int_o^t \lambda\, dt = \int_1^R -\frac{dR}{R} = -\ln R$$

And,
$$R = e^{-\int_0^t \lambda\, dt} \qquad (B.14)$$

Equation B.14 is the most general formulation for reliability. No assumption has been made regarding any specific form for λ and how it varies with time.

Recapping, if we know the failure density function, $f(t)$, we can derive all other reliability functions of interest. We thus see the importance that can be attached either to our ability to experimentally determine $f(t)$, or to credibly assume some form of $f(t)$.

B.3   A Special Case of Interest

The failure density function, $f(t)$, most often used in reliability analyses is the exponential fdf which takes the form:

$$f(t) = \lambda e^{-\lambda t} \qquad (B.15)$$

In Eq. B.15, λ is a *constant* value, and thus for any Δt of interest, λ is a constant. That is to say that the mortality or failure rate is a constant, so the λ in Eq. B.15 is also our λ which we derived as the failure rate.

Or, if you wish, we could assume that λ in Eq. B.14 is a constant and work backwards to obtain $f(t)$:

$$R = e^{-\int_0^t \lambda\, dt} = e^{-\lambda t}$$

$$-\frac{dR}{dt} = -\frac{d}{dt}(e^{-\lambda t}) = \lambda e^{-\lambda t}$$

But,
$$-\frac{dR}{dt} = \frac{dQ}{dt} = f(t)$$

So, when λ is a constant, the corresponding $f(t)$ is:

$$f(t) = \lambda e^{-\lambda t}$$

When λ = constant is assumed (or known), the implications, in hardware terms, are important to understand:

1. The failures in any given interval of time, on average, occur at a constant rate. These failures are random in nature—that is we really don't know just what failure mechanisms are involved or what causes them and, consequently, we do *not* know how to prevent them!

2. If we believe (or know) that λ = constant for the items in question, but we do not know the specific value of λ, we could test 1000 items for 1 hour or a few samples for 1000 hours, and calculate λ. Either way, the resulting values would be approximately the same (if our λ = constant assumption is truly correct).

3. The mean of the exponential fdf, or the mean time to failure (MTTF), is $1/\lambda$. Thus, when the elapsed time of operation is equal to the MTTF:

$$R = e^{-\lambda t} = e^{-\lambda(1/\lambda)} = e^{-1} = 36.8\%$$

4. From 3, we can further understand that when the accumulated operating time is equal to the MTTF, there is a 63.2 percent chance that a randomly selected item in the population has already failed.

Notice that items 3 and 4, when clearly understood, tell us that it is really not wise to use the MTBF as a guide for determining PM task frequency. Among the other pitfalls associated with the λ = constant case, this is one of the unfortunate myths that maintenance organizations often employ.

There is one additional aspect that we should discuss. Every fdf, no matter what form it might take, will have a mean value or, in reliability terms, a mean time to failure (MTTF). In the exponential case, there is an MTTF which we can further label as MTBF due to the $\lambda =$ constant property.

But suppose $\lambda \neq$ constant.

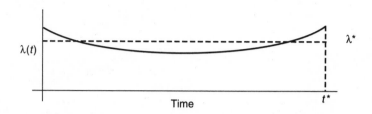

In theory, we can find an average value of λ^* where the area $\lambda^* \cdot t^* = \int_o^{t^*} \lambda(t)dt$. Thus, we could consider λ^* to be a "constant" from $t = 0$ to $t = t^*$, and call $1/\lambda^*$ an average MTBF. In practice, this is frequently done—but without really knowing if we are dealing with a true $\lambda =$ constant case.

This could be dangerous. Consider, for example, a slightly different picture than the one just shown.

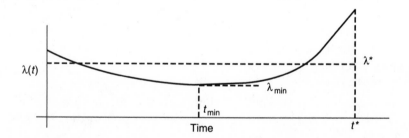

If the equipment operates to t_{min}, then our estimate of λ^* is conservative, and we would experience failure rates *less* than expected. But if we operate to t^*, we would experience failure rates considerably larger than expected (perhaps by 2× or 3×), and this could be very devastating!

Again, the importance of knowing $f(t)$, and whether λ varies with time or not, becomes evident when dealing with equipment and system reliability issues.

References

1. Corio, Marie R., and Costantini, Lynn P., "Frequency and Severity of Forced Outages Immediately Following Planned or Maintenance Outages," Generating Availability Trends Summary Report, North American Electric Reliability Council, May 1989.
2. "RADC Reliability Engineer's Toolkit," Systems Reliability and Engineering Division, Rome Air Development Center, Griffiss AFB, NY 13441, July 1988.
3. *Reliability, Maintainability and Supportability Guidebook,* Society of Automotive Engineers, Inc., Library of Congress Catalog Card No.: 92-60526, ISBN 1-56091-244-8, 2nd Edition, June 1992.
4. Kuehn, Ralph E., "Four Decades of Reliability Experience," *Proceedings of the Annual Reliability & Maintainability Symposium,* Library of Congress Catalog Card No.: 78-132873, ISBN 0-87942-661-6, January 1991.
5. Knight, C. Raymond, "Four Decades of Reliability Progress," *Proceedings of the Annual Reliability & Maintainability Symposium,* Library of Congress Catalog Card No.: 78-132873, ISBN 0-87942-661-6, January 1991.
6. Nowlan, F. Stanley, and Heap, Howard F., *Reliability-Centered Maintenance,* National Technical Information Service, Report No. AD/A066-579, December 29, 1978.
7. Matteson, Thomas D., "The Origins of Reliability-Centered Maintenance," *Proceedings of the 6th International Maintenance Conference,* Institute of Industrial Engineers, October 1989.
8. Personal Communications between A. M. Smith and T. D. Matteson in the period 1982–1985.
9. Bradbury, Scott J., "MSG-3 Revision 1 as Viewed by the Manufacturer (A Cooperative Effort)," *Proceedings of the 6th International Maintenance Conference,* Institute of Industrial Engineers, October 1989.
10. Glenister, R. T., "Maintaining Safety and Reliability in an Efficient Manner," *Proceedings of the 6th International Maintenance Conference,* Institute of Industrial Engineers, October 1989.
11. "Reliability-Centered Maintenance for Aircraft Engines and Equipment," MIL-STD-1843 (USAF), 8 February 1985.
12. *Reliability-Centered Maintenance Handbook,* Department of the Navy, Naval Sea Systems Command, S9081-AB-GIB-010/MAINT, January 1983 (Revised).
13. "Application of Reliability-Centered Maintenance to Component Cooling Water System at Turkey Point Units 3 and 4," Electric Power Research Institute, EPRI Report NP-4271, October 1985.
14. "Use of Reliability-Centered Maintenance for the McGuire Nuclear Station Feedwater System," Electric Power Research Institute, EPRI Report NP-4795, September 1986.
15. "Application of Reliability-Centered Maintenance to San Onofre Units 2 and 3 Auxiliary Feedwater Systems," Electric Power Research Institute, EPRI Report NP-5430, October 1987.
16. Fox, Barry H., Snyder, Melvin G., Smith, Anthony M. (Mac), and Marshall, Robert M., "Experience with the Use of RCM at Three Mile Island," *Proceedings of the 17th Inter-RAM Conference for the Electric Power Industry,* June 1990.

17. Gaertner, John P., "Reliability-Centered Maintenance Applied in the U. S. Commercial Nuclear Power Industry," *Proceedings of the 6th International Maintenance Conference,* Institute of Industrial Engineers, October 1989.

18. Paglia, Alfred M., Barnard, Donald D., and Sonnett, David E., "A Case Study of the RCM Project at V. C. Summer Nuclear Generating Station," *Proceedings of the Inter-RAMQ Conference for the Electric Power Industry,* August 1992.

19. Crellin, G. L., Labott, R. B., and Smith, A. M., "Further Power Plant Application and Experience with Reliability-Centered Maintenance," *Proceedings of the 14th Inter-RAM Conference for the Electric Power Industry,* May 1987.

20. Smith, A. M. (Mac), and Worthy, R. D., "RCM Application to the Air Cooled Condenser System in a Combined Cycle Power Plant," *Proceedings of the Inter-RAMQ Conference for the Electric Power Industry,* August 1992.

21. "Commercial Aviation Experience of Value to the Nuclear Industry," Electric Power Research Institute, EPRI Report NP-3364, January 1984.

22. "RCM Cost-Benefit Evaluation," Electric Power Research Institute, Interim EPRI Report, January 1992.

23. *Reliability-Centered Maintenance (RCM) Technical Handbook,* Electric Power Research Institute, EPRI Report TR-100320, Vol. 1 and 2, January 1992.

24. "A Structured Software Specification for a Reliability-Centered Maintenance (RCM) Workstation," Electric Power Research Institute, EPRI Report NP-7123-2, December 1990.

25. Johnson, Laurence, Colley, Robert, and Morgan, Thomas, "Development of the EPRI Computer Workstation for Reliability-Centered Maintenance Evaluations and Implementation," *Proceedings of the Inter-RAMQ Conference for the Electric Power Industry,* August 1992.

26. EPRI RCM Workstation Software (Release Version 1.0, 6/92), Contact Mr. Thomas Morgan, Halliburton NUS Corporation, Walnut Creek, California.

Index

ABOUT THE AUTHOR

Anthony M. Smith is a senior consultant to government and industry, and is involved in a wide variety of consulting projects in the energy, aerospace, and industrial sectors. During the past 11 years, he has made pioneering contributions to the introduction of the reliability-centered maintenance (RCM) methodology in optimizing preventive maintenance programs at power generating plants. He has more than 37 years of technical and management experience, including 24 years with General Electric. The author of more than 50 technical papers, Mr. Smith is an Associate Fellow of the American Institute of Aeronautics and Astronautics, and a past chair of its System Effectiveness and Safety Technical Committee. He is past General Chairman of the Annual Reliability and Maintainability Symposium, and is currently a member of its Board of Directors. He resides in Saratoga, California, and is a registered Professional Engineer in California.